Barcoding in Mosquitoes

The Authors

Dr. Sathe Tukaram Vithalrao [M.Sc., Ph.D., Sangit Vishard, IBT (Seri.), F.I.S.E.C., F.S.E.Sc., F.S.L.Sc., F.I.C.C.B., F.S.S.I., FHAs] is presently working as Professor and Head, Department of Zoology, Shivaji University, Kolhapur. He has teaching experience of 29 years in Entomology at University PG department and 15 years in Agrochemicals and Pest Management. He has written 30 books and published 255 research papers in national and international journals of repute. He guided 20 Ph.D. students and completed 6 major research projects (from CSIR, DST, DBT and UGC). He visited Canada (1988), Japan (1988), Thailand (2002, 2004), Spain (2005), France (2005), South Korea (2006) and Nepal (2007) etc. for academic work. He is member of editorial board of eleven prestigious journals. He delivered 35 talks through All India Radio and international conferences and involved in Doordarshan, S.T.V. and B.T.V. programmes on useful and harmful insects. He published more than 35 popular articles in daily newspapers on insects and sericulture. He got several prestigious awards like "Environmentalists of the Year-2003", "Bharat Jyoti", "Jewel of India", "International Gold Star", "Eminent Citizen of India", "Education Acumen", "Best Educationist", "Eminent Scientist of the Year-2008", "Lifetime Education Achievement", "Lifetime Achievement in Zoology (Insect Taxonomy)-2009", Education Leadership-2011, Asia Pacific International Award-2012, Global Education Leadership Award-2013, etc. He is also working as Research and Recognition (RR) Committee member for Pune University, Pune; North Maharashtra University, Jalgaon; Shivaji University, Kolhapur and DBA Marathwada University, Aurangabad. He has been awarded several fellowships from different scientific and academic societies. He is Chairman of Maharashtra District Environmental Centre of NESA.

Dr. Mahendra Jagtap (M.Sc., Ph.D.) Entomologist, Malaria Department, District Pune has published 15 research papers and 2 books on mosquitoes. He has participated in several national and international conferences.

Barcoding in Mosquitoes

— Authors —

T.V. Sathe

Head
Department of Zoology
Shivaji University
Kolhapur – 416 004, M.S.

Mahendra Jagtap

Entomologist, Malaria Department
District Pune, M.S.

2014

Daya Publishing House®

A Division of

Astral International Pvt. Ltd.

New Delhi – 110 002

Published by   : **Daya Publishing House®**
                              A Division of
                              Astral International Pvt. Ltd.
                              – ISO 9001:2008 Certified Company –
                              4760-61/23, Ansari Road, Darya Ganj
                              New Delhi-110 002
                              Ph. 011-43549197, 23278134
                              E-mail: info@astralint.com
                              Website: www.astralint.com

Laser Typesetting   : **Classic Computer Services**, Delhi - 110 035

Printed at   : **Replika Press Pvt. Ltd.**

PRINTED IN INDIA

Preface

Molecular techniques have proven very useful for identification of particular genotypes with less man power and time and leading to major medical, veterinary and agricultural advances within the coming years for control of insect pests and insect borne diseases, tuberculosis and HIV/AIDS. Hence in the present text attempts have been made on molecular phylogeny and seasonal abundance of mosquitoes from Western Maharashtra specially from Kolhapur and Sangli districts, collection methods of mosquitoes, DNA and RNAs, history of molecular evolution and DNA sequencing/barcoding of mosquitoes. The book will be helpful to the workers in medical, veterinary, agricultural and biotechnological sciences. The authors are thankful for Madhuri Sathe for her help in completion of this work and Shivaji University Kolhapur for providing facilities. The authors are also thankful to the previous research workers in the molecular phylogeny which stimulated authors for this adventure.

Prof. Dr. T.V. Sathe

Dr. M.B. Jagtap

Contents

1

General Introduction

Insects are one of the most rewarding groups of animals due to their great diversity in form, colour and habitat. They exhibit seasonal and geographical variations and interact with living environment in several ways. However, adverse conditions made by man in the environment such as deforestation, monoculture practices, excessive use of pesticides, habitat modifications etc, lead into biodiversity crisis. These anthropogenic interferences also affect ecology and population dynamics of insects.

The type of insect pest capable of disseminating pathogenic diseases from one mammalian host to another or one plant to another is called vector. Several genera of insects play a role in human disease, but mosquitoes are the most notable disease vectors. The most significant mode of vector borne disease transmission is by biological transmission through blood feeding insects. The pathogen multiplies within the insect vector and the pathogen is transmitted when the insect takes a blood meal. Mechanical transmission of disease agents may also occur when insects physically carry pathogens from one place or host to another, usually on body parts. Majority of vector borne diseases survive in nature by utilizing animals as their vertebrate hosts and are therefore, zoonoses. For a small number of zoonoses, such as malaria and dengue, human is the major host, with no significant animal reservoirs. The key components that determine the occurrence of vector borne diseases refers to-

1. The abundance of vectors and intermediate and reservoir hosts.
2. The prevalence of disease causing pathogens suitably adapted to the vectors and the human or animal hosts.
3. The local environmental conditions especially temperature and humidity.
4. The resilience behaviour and immune status of human population.

Dipterans dominate as vectors of a number of human and livestock diseases over other members of the orders. Mosquitoes profoundly play an important role making enormous impact by spreading epidemics. The disease outbreaks are due to both biological and environmental factors that encourage vector breeding. Therefore, to study the relationship between humans, vectors and pathogens with respect to disease transmission is very crucial.

The mosquitoes belong to the phylum Arthropoda, Order Diptera, Sub Order Nematocera and the family Culicidae. According to Knight and Stone (1977) there are 3220 species of mosquitoes in the world belonging to 34 genera, out of India represents 16 genera (Barraud 1934).

Mosquitoes are small, slender, two winged insects covered with hairs and scales, and with three body divisions *viz.* head, thorax and abdomen. Prognathus head contains a pair of compound eyes and densely hairy with 14-15 segmented antennae, plumose in males and pilose in the females. Proboscis is long, brown or black piercing and sucking blood. Palpus is about the length of the proboscis in the males but, in females it differs in length in the different genera. Thorax is convex, made up of pro, meso and metathorax and externally undifferentiated. It consists of pair of wings and pair of halters. Thorax carries a three pairs of legs. Each leg consists of coxa, followed by femur, tibia and tarsus. Tarsus is composed of five joints. Each leg ends with a pair of claw. Abdomen, the last division of body consists of eight segments and terminates in a pair of claspers in males and in the female as lobed appendages.

Primarily males and females of mosquito species feed on nector and other plant juices. However, the females of some species need vertebrate blood to produce eggs. The females thus have very specialized long, piercing and sucking mouth parts. By piercing the skin of their prey they obtain the blood feed. Individual female mosquitoes can lay eggs in several batches in their life span. However, a blood meal is essential for each batch. The females of different species feed at different times of the day and in different types of locations within the same surroundings. Any vertebrate animal may provide the required proteins to mosquitoes. The mosquitoes have specific prey feeding preferences.

The life cycle of mosquitoes passes through four distinct stages *viz.* egg, larva, pupa and adult (Busvine, 1980). Eggs are laid either singly or in clusters. The cluster may contain about 300 eggs attached together to form a soft of 'raft'. Some mosquito species lay eggs on the surface of calm water, while others lay on damp soil destined to be flooded either way, the eggs end up floating at the water's surface. Incubation period is 2-3 days. In some cases eggs laid in late season may withstand the harsh conditions of winter and hatch in the following spring (Busvine, 1980).

Larval period of mosquito varies from 4-14 days in the water. They breathe air through a tube to the surface and eat microorganisms and other organic matter from water. The larvae metamorphosed into pupae (tumblers). The pupal stage is non feeding stage. Pupae float at the surface but are capable of diving/tumbling to safety when disturbed. The pupal period is about 1-4 days. Fully developed adult mosquito breaks the pupal case and comes out to the surface of the water, where it rests until its body dries and hardens. After two days of emergence mating and feeding occurred.

Temperature affects species specific characteristics and the rate of development. Under favourable warm conditions, some mosquitoes develop from egg to adults in less than a week. Adult female mosquitoes mostly survived for about two weeks. They may parasitize humans or predated by birds, bats, amphibians or spiders. However, adult females may hibernate throughout the winter if emerge late in the season. They start egg laying in the spring.

Mosquitoes are responsible for spreading various fatal diseases like malaria, filariasis, dengue, chikungunya, encephalitis, yellow fever etc.

Fowl pox of poultry, mycomatosis of rabbits, rift valley fever of sheep, encephalitis of horses and birds, heat worm of dogs etc are also transmitted from animal to man and some diseases of from other insects. South American warble fly *Dermatobia* sp. transports its eggs to the skin of man and animals through mosquitoes alone and cause myiasis on skin after hatching (Sathe and Tingare, 2010).

The human malarial protozoan pathogen refers to *Plasmodium vivax, Plasmodium falciparum* (malignant tertian, subtertian aestivoautumnal malaria, falciparum malaria), *Plasmodium malariae* (quartanmalaria) and *Plasmodium ovale* (ovale malaria). Races and strains are visualized in case of *P. vivax* and *P. falciparum*. They are confirmed through the clinical picture, geographical distribution and immunological responses. *P. vivax* causes 65-90 per cent malaria while, *P. falciparum* 25-30 per cent of the total malarial cases (Sathe and Girhe, 2001). *P. malariae* shows only 1 per cent cases but, *P. ovale* is reported sporadically. The protozoa invade the parenchyma cells of the liver and after passing through development stage, attack and resides inside RBCs. Female mosquito inject saliva and sporozoites during the act of biting and feeding. Other means of malarial transmission can be either through blood transfusion or congenital transmission as the use of a syringe from an infected person. (Sathe and Tingare, 2010).

The species of the genus *Anopheles* specially *Anopheles labranchiae* Flalleroni, *An. sacharovi* Favre, *An. sargenti* (Theobald), *An. superpictus* Grassi and *An. pharoensis* Theobald are reported from parts of the Mediterranean area, *An. funestus* Giles, *An. moucheti* Evans, *An. nili* (Theobald) and member of the *An. gambie* Giles complex in the Ethiopian region, *An. stephensi* Liston, *An. fluviatilis* James and *An. pulcherrimus* Theobald from Western Asia *An. culicifacies* Giles from India and Ceylon, while, *An. maculatus* Theobald, *An. sundaicus* (Rodenwaldt) and member of the *An. barbirostris* Vander Wulp, *An. hyrcunus* (Pallas) Wulp group are reported from Southern Asia. The members from *An. punctulatus* complex in Melanesoa and *An. pseudopunctipennis* Theobald, *An. bellator* Dyar and Knab, *An. cruzi* Dyar and Knab, *An. darlingi* Roof, *An. aquasalis* Curry, *An. albimanus* Wiedemann, *An. albitarsis* Lynch Arribalzga and *An. nuneztovavi* Gobaldon are reported from Central and South America (Smith, 1969).

The Indian species of *Anopheles* transmitting malaria refer to *An. annularis* Van der Wulp, *An. sundaicus* RodenWaldt, *An. stephensi* Liston, *An. maculatus* Theobald, *An. philippinensis* Ludlow, *An. leucosphyrus* Doenitz., *An. fluviatilis* James, *An. culicifacies* Giles, *An. varuna* Iyengar and *An. minimus* Theobald (Rao, 1984). The dominant species of *Culex* includes - *Culex pipiens* Linnaeus, *C. vishnui* Theobald, *C. pseudovishnui* Colless, *C. tritaeniorhynchus* Giles, *C. bitaeniorhynchus* Giles, *C. sinensis* Theobald, *C. gelidus* Theobald, *C. sitiens* Wiedemann while, the *Aedes is* represented by *Aedes aegypti* Linnaeus, *A. albopictus* Skuse, *A. vittatus* Bigot and *A. variegatus* Schrank.

Filaria pathogens in man are two nematodes viz, *Wuchereria bancrofti* and *Brugia malayi*. The adult worms live in the lymphatics, produce live embryos (microfilariae) which invade the blood stream. The embryos do not develop further in the blood. According to Sathe and Tingare (2010) a mosquito vector imbibes microfilariae along with the blood and develop further in 10-12 days, depending upon the environment. The third stage infective larvae are deposited on the human skin by the mosquito when it visit man for the next food. Such larvae enter through the wound. The minimum time taken from the entry of the third stage infective larvae until the microfilariae first appear in the peripheral blood is about 82 days, but it may take as long as one year. The female nematodes begin to liberate microfilariae which invade the blood stream. The cycle is thus repeated.

It is estimated that at least 250 million people are infected with *W. bancrofti* and *B. malayi* from the world. About one billion people from tropical and sub-tropical countries are at risk and about 200 million are actually suffering from the disease. In India more than 200 million populations is at risk. From the world about 90 species of mosquitoes have been recorded as natural vectors of *W. bancrofti*. *Culex pipiens* Linnaeus is the vector in India. *Anopheles gambie* A and B and *Anopheles funestus* are important vectors of filaria in rural East Africa.

Coastal parts and banks of rivers are distributional places of *Bancroftian filariasis* in South India. The most favourable period of transmission is during the monsoon but found throughout the year. The favourable conditions for vector and parasite are 15.5 °C–32.2°C temperature and at least 60 per cent relative humidity (RH).

Brugia malayi cause Malayian filariasis from Orissa, Madhya Pradesh, Tamil Nadu, Kerala, West Bengal and Assam. The mosquitoes of the *Mansonia* group and *Anopheles barbirostris* are the vectors in India. Important reservoirs for this worm are monkeys (*Pereshytis obscurus*), cat, dog and some other animals.

JE (Japanese Encephalitis) is transmitted by *Culex* mosquitoes. The pathogen involved is Virus (JEV) which is maintained in nature by a complex cycle that involves pig as amplifying host arderid birds as reservoirs and mosquitoes as vectors. *Culex vishnui, Culex tritaeniorhynchus* Giles, and *Culex pseudovishnui* Colless have been implicated as major vector of JE in India. 16 species of mosquitoes are susceptible for JEV. These mosquitoes transmit viral encephalitis in man. JE is an arboviral (B) infection. The virus was first isolated in 1935 by Japanese workers.

Yellow fever is transmitted by *Aedes* and other mosquitoes is an acute specific viral fever of short duration. YF virus is found in certain wild monkeys and other reservoir hosts in Africa and South America. *A. aegypti* mosquitoes bite infected men and led mosquito-man cycle. In South America, daybiting mosquitoes *Haemagogus spegazzinii, Aedes leucocelaneus* and *Sabethes chloropterus* are vectors and monkey-mosquito monkey cycle goes on. In West Africa, night biting mosquitoes *Aedes, Culex, Ochlerotatus, Sabethes* and *Anopheles* transmit yellow fever virus to human beings in the forests. Infected monkeys may enter human habitations and became source of further transmission. Incubation period in man is 3-6 days (Atwal, 1933).

Dengue Fever (DF) is an acute viral infection for which *Aedes aegypti* is vector. It breeds in clean water containers and elsewhere. Rush, in 1780, called it as 'break-bone' fever. Graham in 1905 was the first to report the mosquito vector for dengue. *A. aegypti* as a vector of dengue was proved in 1906 by Bancrofti in Australia. The arbovirus (B group) is existing in four forms *viz.*, Dengue 1, Dengue 2, Dengue 3 and Dengue 4, all are transmitted mainly by *A. aegypti*. Other species of the genus *Aedes* (*A. albopictus, A. squtellaris, A. albimanus* and *A. hebridus*) and *Armigeres obtarbans* also known as vector for Dengue. *A. aegypti* maintains the disease cycle in man, while *A. albopictus* and others living in the bush or forests help in maintenance of infection among monkeys (jungle dengue). The Dengue is prevalent in tropical and sub-tropical areas of the world.

The most troublesome disease, chikungunya was first reported from Kolkatta in 1963. Its name is derived from the swahili word meaning "that which bends up" in reference to stooped posture developed as a result of the arthraligia (severe joint pains). The first outbreak of CHK virus in Kolkatta followed by Chennai, Pondichery, Vollore and Vishakapattanam in 1964. Later, it was recorded from Central part of India. *i.e.* Rajmundri, Kakinada (A.P.) and Nagpur in 1965. Chikungunya is high as 40 to 70 per cent in certain wards of Nagpur in 1965. Now all the age groups are susceptible to CHK and is widely distributed throughout India.

The virus is maintained in nature at a low level in man-mosquito-man cycle. The survival of CHK virus in nature is also through transovarial transmission (TOT) in *A. aegypti* mosquitoes. *A. aegypti* is the principal vector of this virus in India. However, chikungunya can be transmitted by *Aedes albopictus, Aedes vittatus* and some *Culex* species.

When a mosquito bites to west nile virus infected bird, it picks up viral particles circulating in the blood. Once inside the body of a mosquito they can serve as a vector, the virus co-opts the mosquito's cells into replicating the virus. After 5-14 days virus multiplies and goes to salivary glands of mosquito from where they are transported into the subsequent host when the mosquito bites for meal. West Nile Virus is primarily an avian pathogen and is transmitted among birds by ornithophilic (bird biting) mosquitoes.

Study Area

Western Maharashtra (Figure 1) is leading in agriculture and industrialization and have several water bodies and hilly areas like Western Ghats. Due to Western Ghats, the geography and the climate of Western Maharashtra is more or less similar in the districts Kolhapur, Satara, Sangli and Pune (Figures 2 and 3), vary from place to place. The districts selected for the study are shown in Figures 2 and 3. Hence, Western Maharashtra have been selected for the present work.

Kolhapur district (Figures 2 and 3) is located between 15° to 17° North latitude and 73° to 74° East longitude. The district is bounded by Sangli district at the North, Belgaum district of Karnataka State at the South and East and Ratnagiri and Sindhudurg districts at East and West respectively.

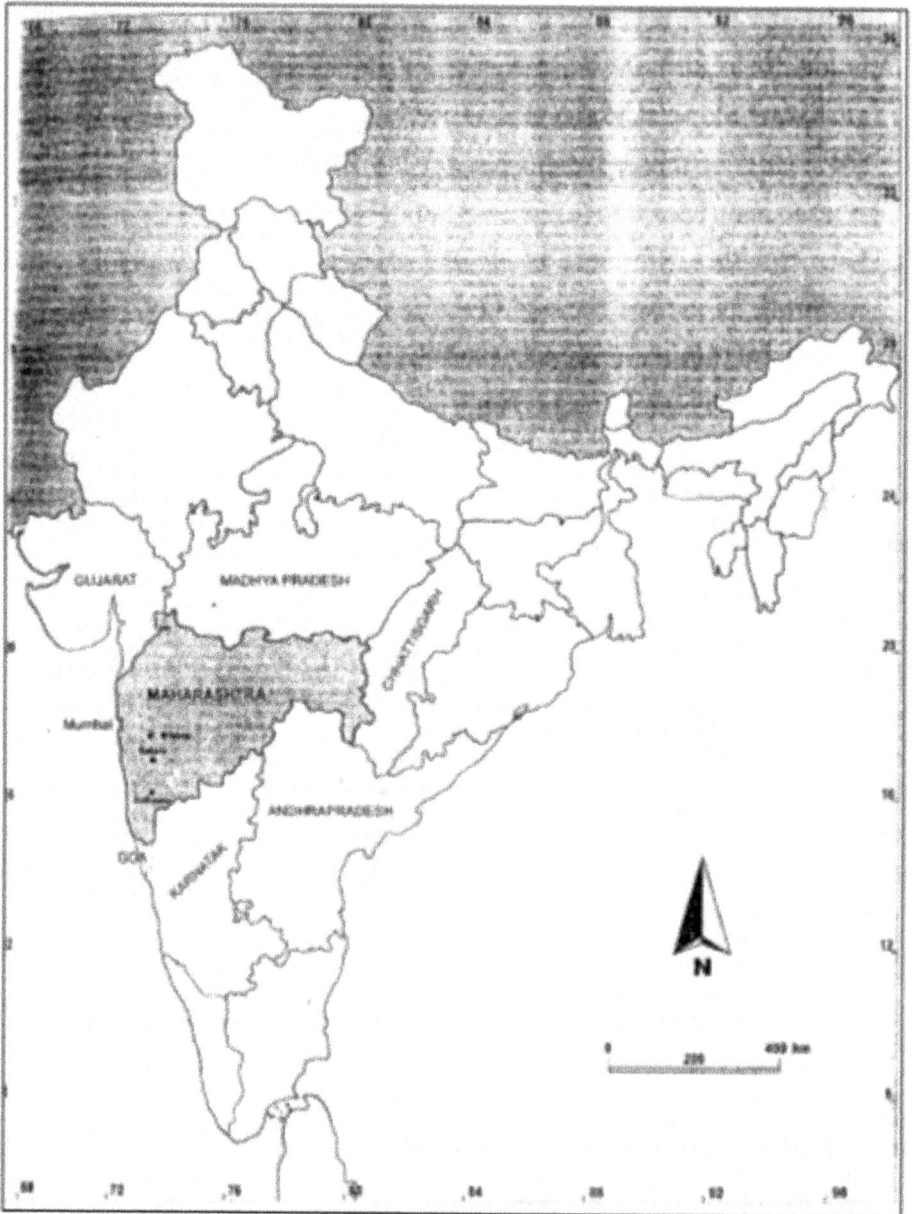

Figure 1: Map of India Showing Maharashtra.

Kolhapur district has 7633 sq.km area with population more than 20,03,953. 12 tahsils are included under this district.

Sahyadri mountains in the west, provides several spur's in the East of the district. Major portion of the district is 390 to 600 metres above mean sea level. Krishna, Warna, Panchaganga, Doodhganga, Vedganga and

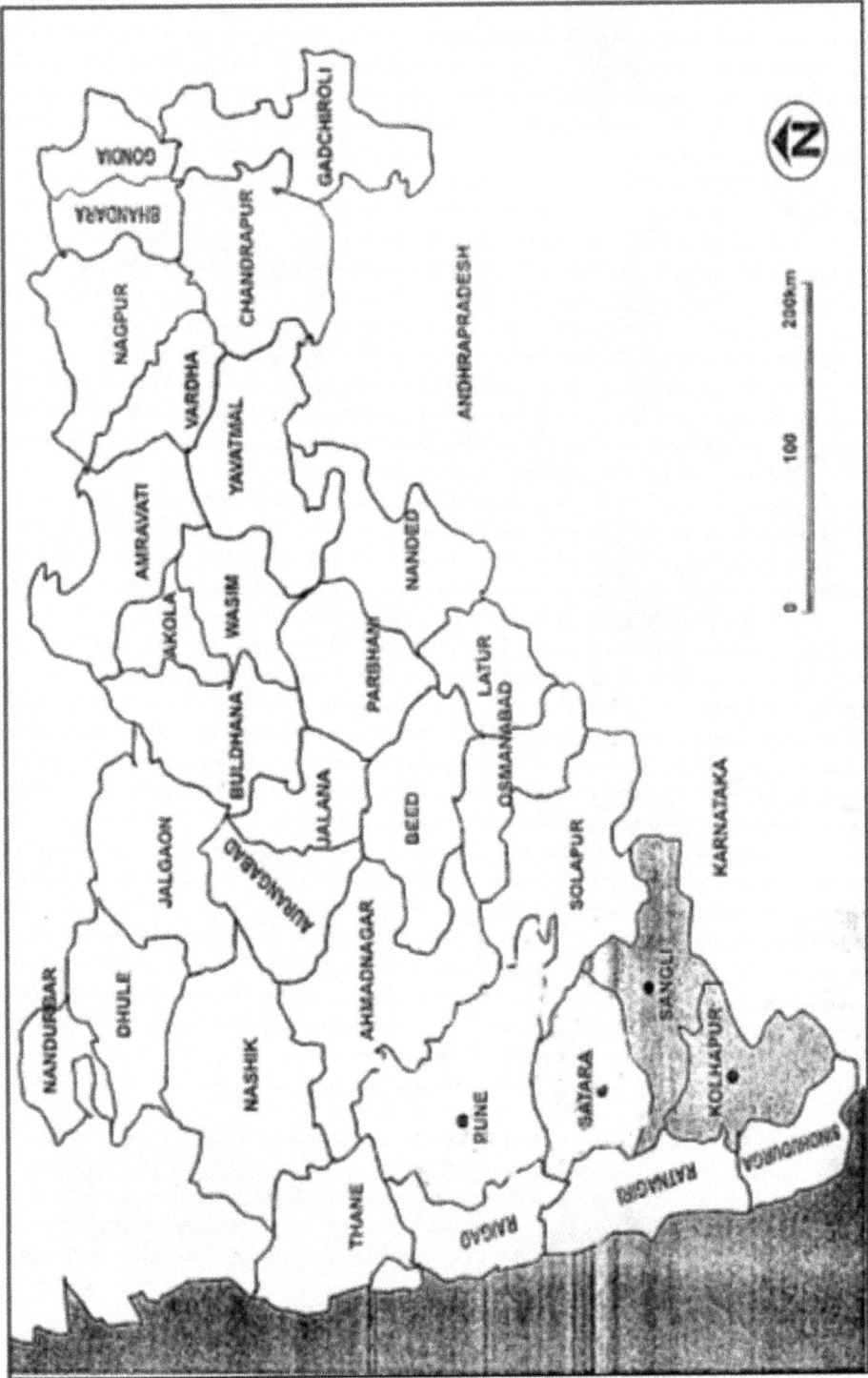

Figure 2: Map of Maharashtra Showing Study Area (Western Maharashtra Districts Kolhapur, Sangli, Satara and Pune.

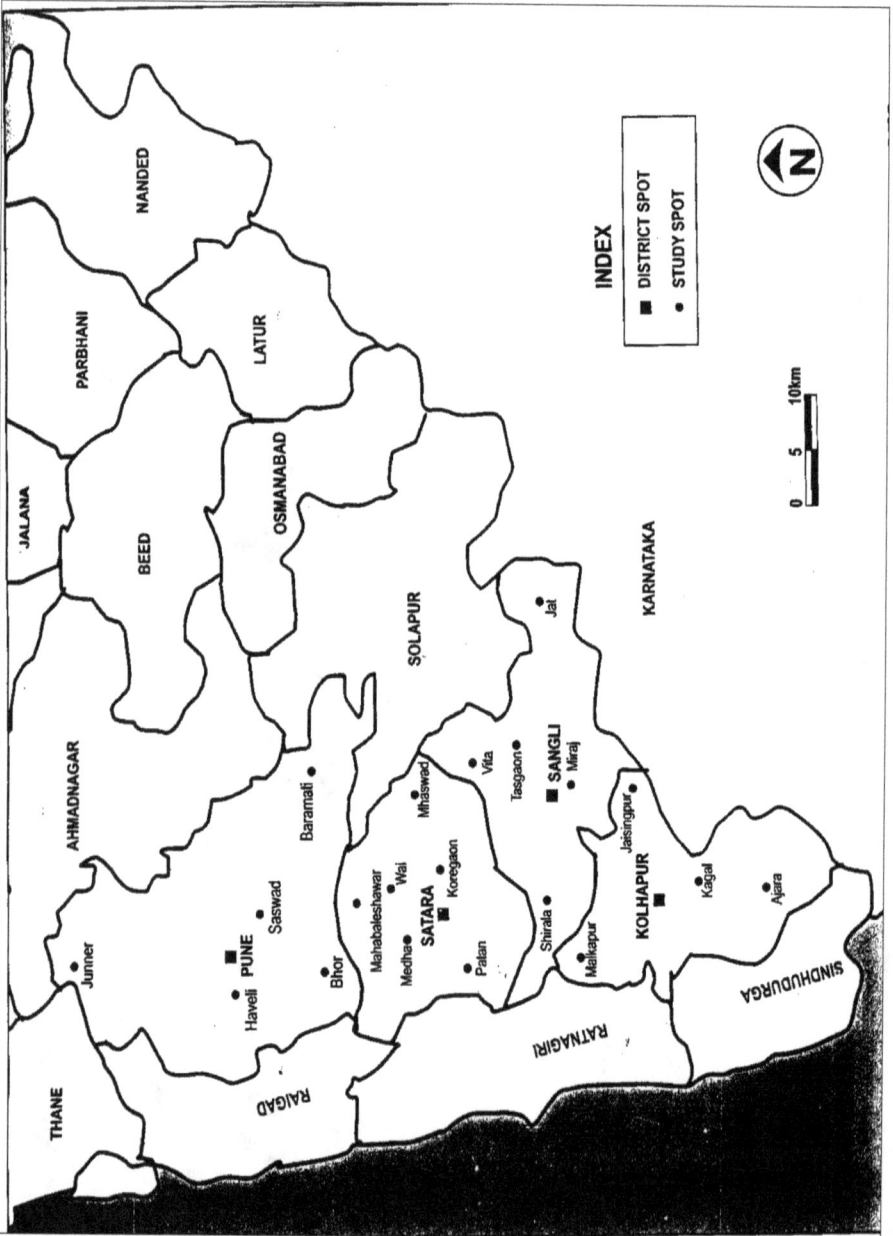

Figure 3: Map of Western Maharashtra Showing Districts and Study Spots.

Hiranykeshi are the principal rivers of Kolhapur district. Panchganga is formed by the five tributaries namely, Kasari, Kumbhi, Dhamana, Tulashi and Bhogawati. Panchganga merge into the Krishna at Narsobawadi in Shirol tahsil. The South Western region of the district is drained by Doodhganga river. Forests in Kolhapur district are confined to the Western half of the district. The total forest area in Kolhapur district is more than 1,46,575 hactares. The rainfall is not evenly distributed in the district. Gagan Bavada receives over 6000 mm rainfall while, Hatkanangale in the East receives little as 500 mm. The district gets rain from the South West as well as from the South East monsoon and the rainy season is from June to November, the above facts clearly indicates that Kolhapur district is good place for mosquitoes.

Ajra, Chandgad, Bavada, Radhanagari and Shahuwadi tahsils come in the heavy rainfall tract (2096-6232 mm) while, Hatkanangale and Shirol tahsils are under poor rainfall tract (600 mm). Tanks/dams namely, Kalamawadi dam, Radhanagari dam, Kalamba tank, Rajaram tank, Rankala and several others are good places for mosquito breeding. From Kolhapur district five spots were selected namely, Kolhapur, Jaysingpur, Kagal, Malkapur and Ajara.

The Sangli district (Figures 2 and 3) lies on the river basins of Warna and Krishna. Sangli district lies between 16°.45' and 17°.38' North latitudes and 73°.42' and 75°.40' East longitude. Sangli district is located towards the eastern part of the state of Maharashtra surrounded by Satara, Solapur districts to the north, Vijapur district to the east, Kolhapur and Belgum districts to the south and Ratnagiri district to the west. Area of the district lies partly in Krishna basin and partly in Bhima basin. The maximum temperature ranges between 31.1°C to 41.5°C. Similarly, the minimum temperature ranges from 10.3°C to 21.5°C. In Sangli district 76 major and minor irrigation projects have been launched. Therefore, mosquitogenic conditions are favourable for transmitting the vector borne diseases. From Sangli district five spots were selected namely, Miraj, Jath, Tasgaon, Vita and Shirala.

Kolhapur and Sangli districts have been selected from Western Maharashtra as a study area. On the basis of geographical and climatical parameters these districts were selected. Secondly, these districts have great importance in agriculture and industrialization and thirdly they contain several breeding places for mosquitoes.

2

Collection and Preservation of Mosquitoes

Materials and Methods

Correct identification of disease vector is important for studying its useful and harmful aspects. Vector species may be sampled to assess presence/absence or abundance in order to determine whether control measures are necessary. Following material and methods were adopted in completion of the work since any minor change can affect the results.

Materials

Mosquitoes resting on different surfaces (indoor or outdoor) were collected by hand collection method by using suction tube and insect net.

1) Suction Tube (Figure 4)

The suction tube contains a glass or plastic tube of 15 mm diameter and 20-30 cm length; a flexible rubber or plastic tube, 80-100 cm long and 15-18 mm in diameter, mesh between glass tube and rubber tube for preventing entry of mosquitoes and solid dusts towards the mouth. The third portion of the suction tube is mouthpiece for sucking mosquitoes by inhalating air through mouth. After sucking five mosquitoes, the sucked specimens were separated in the test tubes/specimen bottles.

2) Test Tubes (Figure 5)

Two types of test tubes, one with open ends of both sides and another only one end open were used for mosquito transfer and storage (15 × 15 mm and 15 × 25 mm length and diameter). During the handling of adult mosquitoes the test tubes were plugged with cotton at the mouth or both the ends in case of open tubes.

3) Torch (Figure 6)

The Torch is essential for the adult surveillance and identification of resting mosquitoes in various habitats in day or night.

4) Mosquito Rearing Cage (Figure 7)

Mosquito larvae collected from field spots in beakers were reared in the cage (size of the cages was 25 × 25 × 25 cm) Three sides contain nylon mesh and at one side, with muslin cloth sleeve as door for easy handling of mosquitoes.

5) Specimen Tubes (Figure 8)

Specimen tubes of size, 6 × 2 cm, 5 × 2 cm and 4 × 2 cm (length and diameter) were used for preserving and handling the adult mosquitoes. The open end of specimen tube was plugged either by rubber or wooden cork/cotton balls. Specimen tubes were used for keeping pinned adult mosquitoes.

6) Insect Net (Figure 9)

In outdoor and indoor, mosquitoes were collected by hand net. Insect hand net contain aluminium handle of 70 cm long, circular iron ring of 22 cm diameter and ordinary mosquito nylon mesh bag of 70 cm depth.

7) Dipper (Figure 10)

The dipper (Figure 10) was used for collecting and handling mosquito larvae and pupae from natural habitat. The larvae and pupae will remain in the dipper and the water will be dropped down.

8) Ladle (Figure 11)

Ladle is used for collecting mosquito larvae and pupae from various aquatic habitats of mosquitoes. It consists of a flat handle of length 25 cm with a circular concave cap of diameter 8 cm (Figure 11).

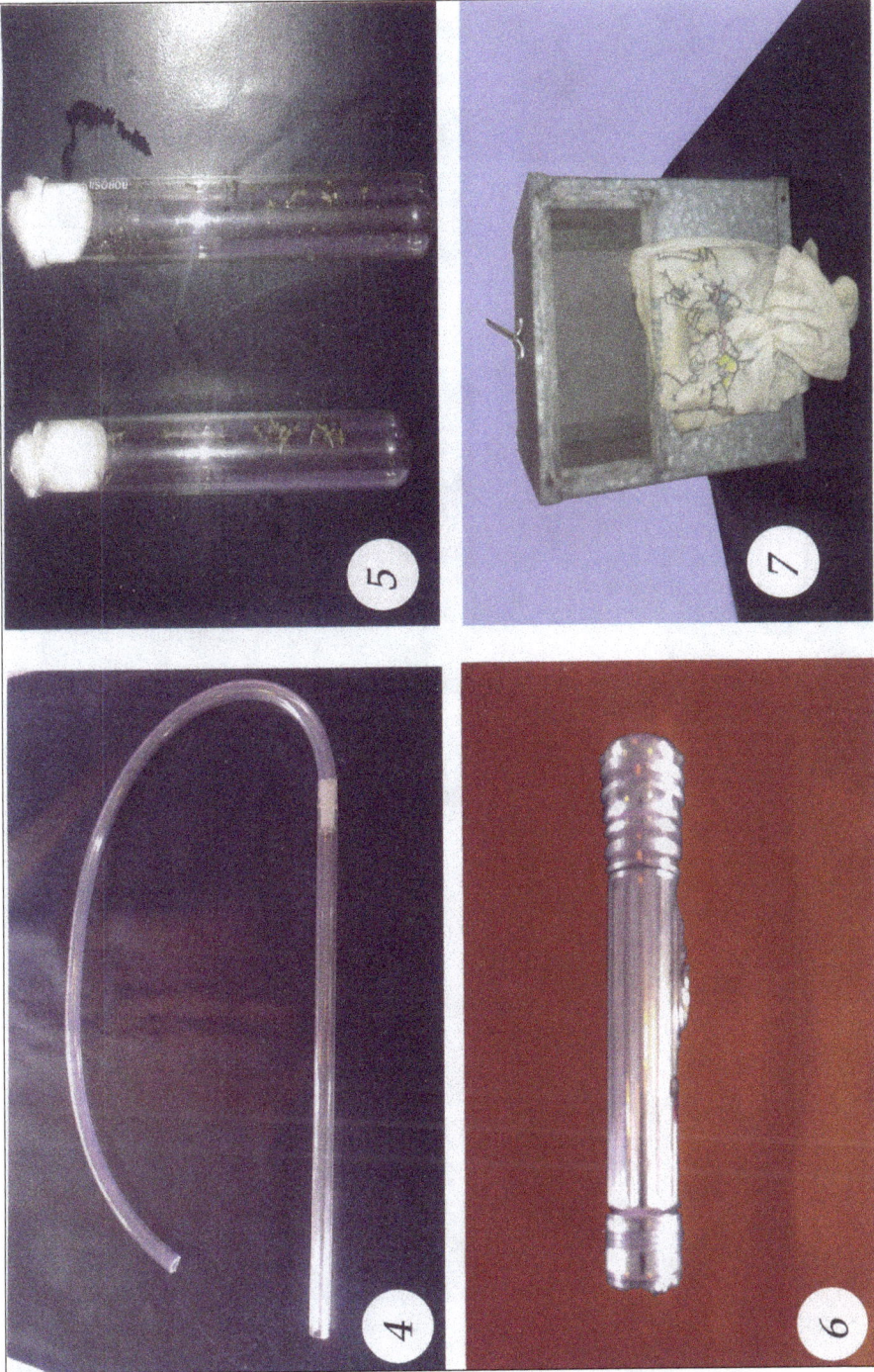

Figure 4: Suction tube; **Figure 5:** Test tubes; **Figure 6:** Torch; **Figure 7:** Mosquito rearing cage.

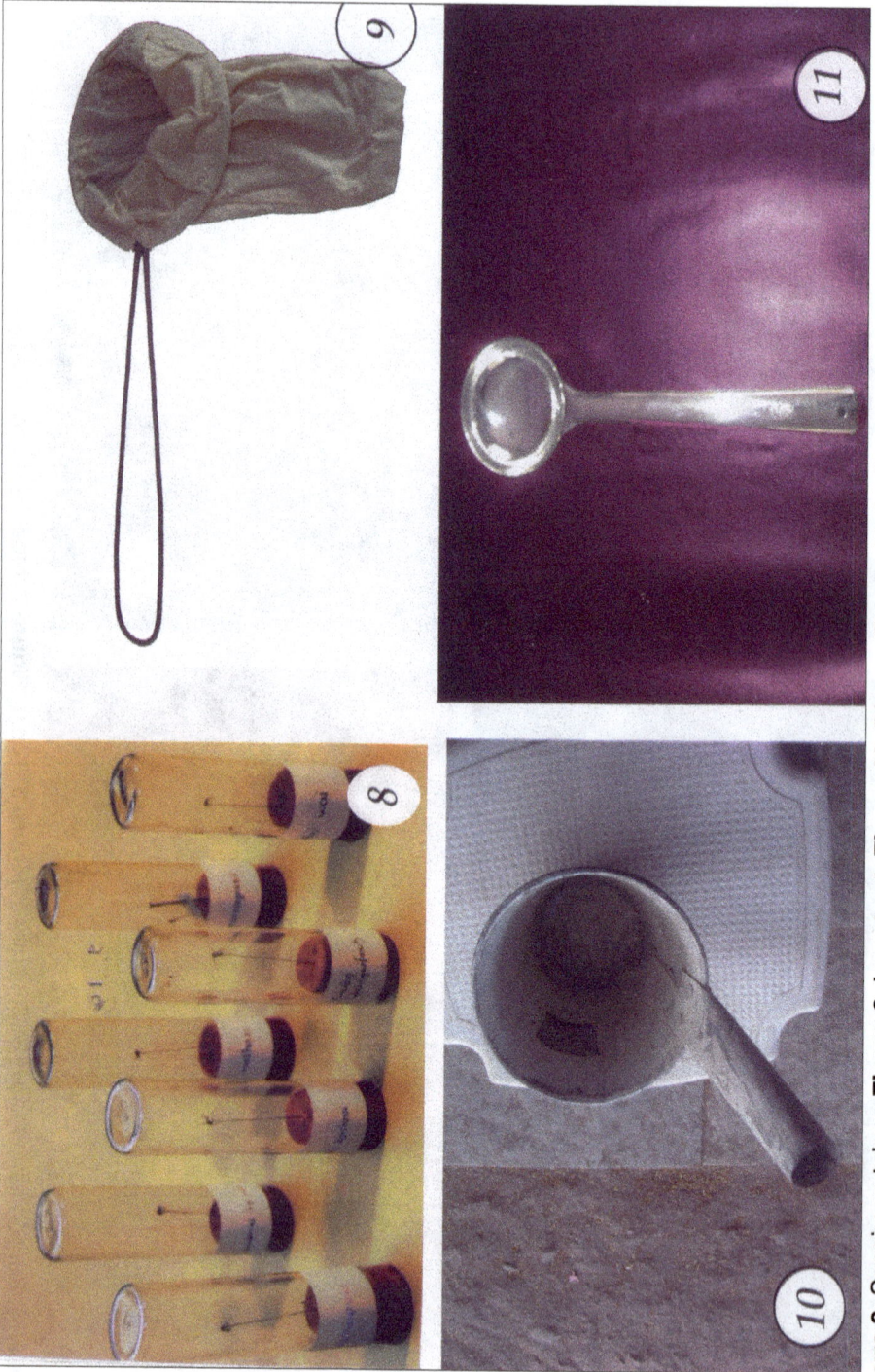

Figure 8: Specimen tubes; **Figure 9**: Insect net; **Figure 10**: Dipper; **Figure 11**: Ladle.

9) Plastic Containers (Figure 12)

The larvae and pupae collected must arrive alive and undamaged at the laboratory. Hence, plastic containers were used for collecting the samples of larvae/pupae from the natural habitat.

10) Rubber Droppers (Figure 13)

Rubber droppers were used for picking the larvae/pupae from natural habitat and handling them in laboratory.

11) Specimen Bottles (Figure 14)

The adults and larvae were preserved in specimen bottles (size, 6 × 2 cm, 5 × 2 cm and 4 × 2 cm length and diameter). The mouth of bottle was closed by rubber cork/plastic cap for keeping specimen air tight and safety.

12) Camel Hair Brushes (Figure 15)

Camel hair brush No. 1 and 2 has been used for handling the specimen and preparation of slides.

13) Camera Nikon S4 (Figure 16)

Nikon Coolpix S4 camera, 6 megapixal and 10x optical zoom was used for photography of the mosquitoes and mosquito habitat.

14) Slide Box (Figure 17)

Slide box was used for keeping the permanent slides safely. Slide boxes of size, 28 × 22 × 3.5 cm, 21 × 19 × 3.5 cm (length, width and height) were used.

15) Slides and Cover Slips (Figure 18)

Ordinary slides and cover slips were used for preparing the whole mounts and other body parts such as head, proboscis, antenna, thorax, wing, halter, hind leg and abdomen.

16) Oven

Oven of size, 3.6 × 2.4 feet (height and width) has been used for drying adult mosquitoes and the permanent slides of mosquitoes.

17) Compound Microscope (Figure 19)

Simple monocular compound microscopes with objectives 10x, 45x, 100x was used for describing the mosquito species with the help of occulometer.

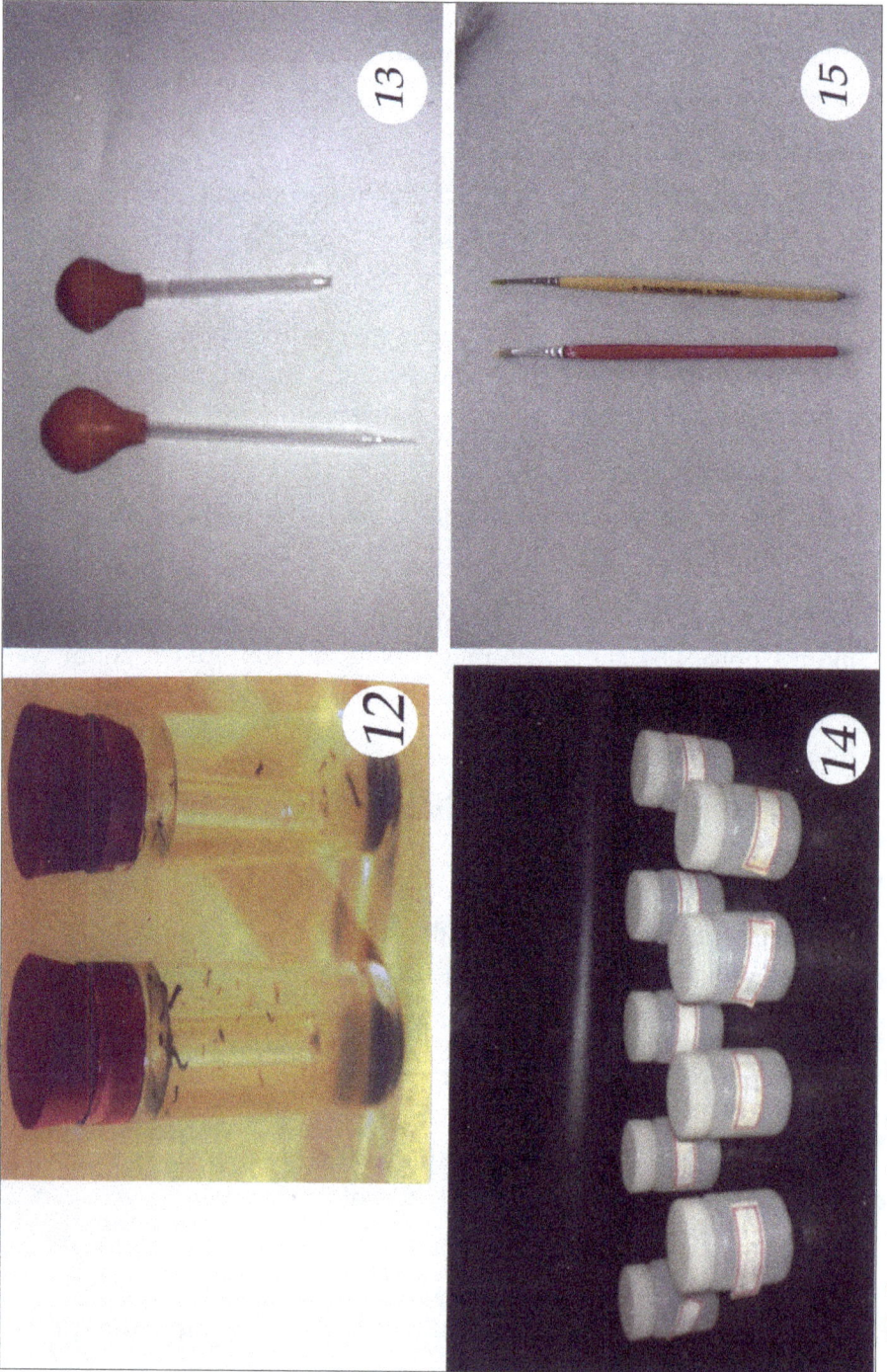

Figure 12: Specimen bottles; **Figure 13**: Dropper; **Figure 14**: Plastic containers; **Figure 15**: Camel brush.

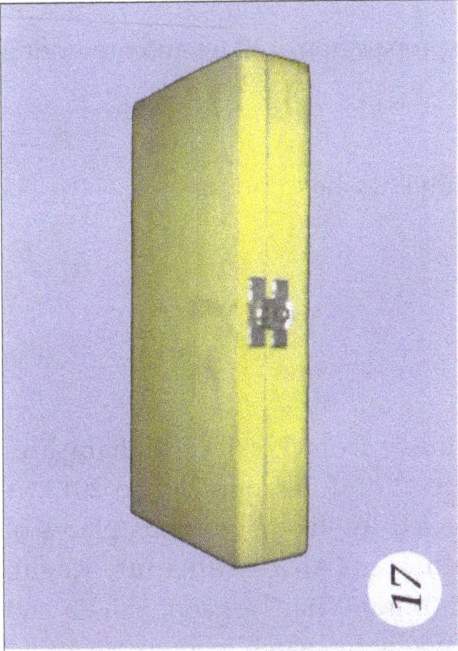

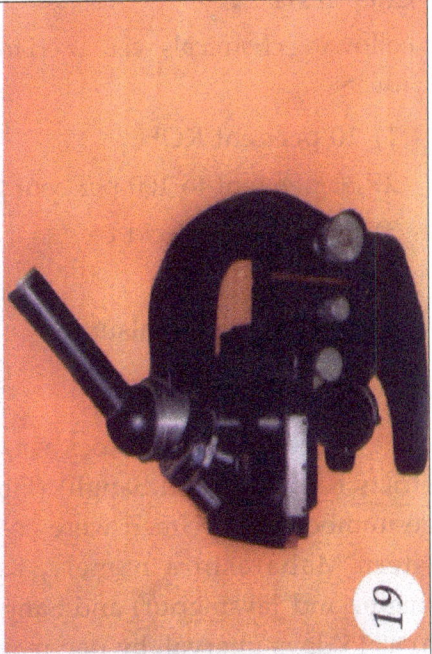

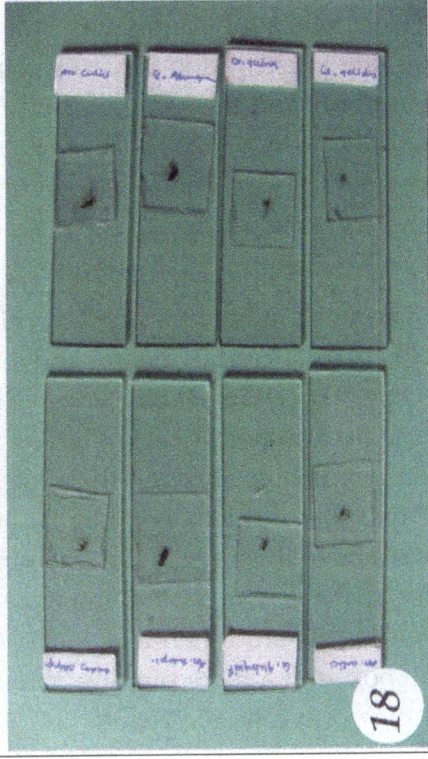

Figure 16: Camera; **Figure 17**: Slide box; **Figure 18**: Slides and coverslips; **Figure 19**: Microscope.

18) Chemicals

Following chemicals were used for preparation of slides and preserving the insects.

1) 10 per cent KOH.

2) 30 per cent to 100 per cent Ethyl alcohol grades.

3) Glacial acetic acid etc.

4) Xylene.

5) DPX/Canada Balsum.

Methods

The survey of mosquitoes was made from Western Maharashtra (districts Kolhapur and Sangli) (Figure 2 and 3) from 2007 to 2013. A large number of specimen were collected by visiting various places of Western Maharashtra namely, Kolhapur (Ajra, Malkapur, Kagal, Kolhapur and Jaysingpur) and Sangli (Miraj, Vita, Tasgaon, Shirala and Jath) at 15 days interval, by one man one hour search method.

Most commonly, a mixture of 75 per cent alcohol to 25 per cent water is used for preservation of insects. The water should be distilled to ensure a neutral pH and the solution should be thoroughly mixed since alcohols and water do not mix easily by themselves. The adult mosquitoes were narcotized in ether and killed in killing bottle by chloroform. Specimens for molecular work should be collected in 95 per cent or absolute (100 per cent) ethanol (ethyl alcohol).

The mosquitoes were pinned with entomological pins from the ventral side, kept on spreading board, and dried in drying chamber/oven at 60°C. The dried specimen then kept in Sterilized specimen tube by pinning invertedly to wooden cork. Pinning refers to the insertion of a standard insect pin directly through the body of an insect from ventral side using care that the pin does not tear off any legs or destroy chetotaxy on the dorsal side of the insect. After the pin is inserted and before the specimen is dry, the legs, wings, and antennae should be arranged so that all parts are visible for study. Pinned specimens should always be placed as in a small box with a foam pinning bottom. The box should be well wrapped and placed in a larger carton.

Specimens are mounted so that they may be handled and examined with the greatest convenience and with the least possible damage. Well-mounted specimens enhance the value of a collection; their value for research may depend to a great extent on how well they are prepared and preserved. For the taxonomical study, head, antenna, proboscis, wing, legs, halter and abdomen were mounted on slide in D.P.X. Specimen have been prepared on the slide and the slides are kept in slide boxes. The final stage in preparing permanent mounts is thorough drying or hardening of the medium. This may be done in any clean environment or in an oven or special slide warmer under gentle heat. The mounts should be carefully labeled either before drying or afterward. The records were made on locality, date of collection and identification. To label microscope slides, square labels are used. Morphological studies were carried out with the help of monocular microscope. Comparative measurements of body parts of specimen were made with ocular micrometer and calculated with the help of graduated mechanical stage. All measurements were made in millimeter.

The terminology adopted for description of species is same as that of Christophers (1933), Barraud (1934), Puri (1948), Horsefall (1955), Knight and Stone (1977), Tanaka (1979), Rao (1984), Nagpal and Sharma (1995), Sathe and Girhe (2002), Sathe and Tingare (2010) and Sathe and Jagtap (2013). A large number of references were consulted in the course of the study and cited in the text.

Molecular Phylogeny

Mosquito specimens used for constructing DNA barcodes were from collections made from different study spots. Genomic DNA extraction was carried out from whole mosquito using the QIAamp DNA Mini Kit (QIAGEN) method. A neighbourhood joining (NJ) tree of K2P distances was created to provide a graphic representation of the clustering pattern among different species (Saitou and Nei 1987, Hajibabaei *et al.*, 2006). These analyses of the sequences were conducted using MEGA version 4 software (Kumar *et al.*, 2004). The holotype and paratypes are time being with Department of Zoology and will be deposited at ZSI, Kolkatta in due course time. Detail methodology of phylogeny is given in chapter 5.

3

Seasonal Abundance and Distribution of Mosquitoes

Introduction

Knowledge upgradation is important to monitor the impact of rapidly changing ecological conditions such as deforestation, population movement and developmental activities on mosquito distribution and vector bionomics (Rahman *et al.*, 1977). Seasonal abundance of mosquitoes may vary spatially. Sampling of mosquito population is an important task, which estimate the number of species presents in a target area. Patterns of seasonal abundance of certain mosquito species are correlated to proliferation of its breeding habitats during rainy season and its scarcity during dry season (White, 1974). Various populations have specific characteristic features which facilitate the formation of epidemiological characteristics of vector borne diseases (Kondrashin and Kalra, 1987). Rapidly changing environment brings about frequent changes in vector behaviour, which affects the vector bionomics (Prakash *et al.*, 1998).

Global warming is reshaping the ecology of many medically important insect vectors. Warmer temperatures have been shown to directly increase mosquito biting and pathogen transmission.

Perusal of literature indicates that several workers (Senior White, 1937; Senior White *et al.*, 1943; Foot and Cook, 1959; Nagpal *et al.*, 1983; Das *et al.*, 1984; Rao, 1984; Nagpal and Sharma, 1987; Nagpal and Sharma, 1995; Reuben *et al.*, 1992; Rajavel *et al.*, 2000; Sathe and Girhe, 2001a, 2001b, 2001c; Murty *et al.*, 2002; Kanojia *et al.*, 2003; Sharma *et al.*, 2005; Joshi *et al.*, 2005; Tilak *et al.*, 2006; Pemola and Jauhari, 2006; Malarial Research Centre, 2006; Baruah *et al.*, 2007; Jagtap and Sathe, 2008a; Jagtap and Sathe, 2008b; Jagtap and Sathe, 2008c; Sathe and Jagtap, 2009; Jagtap and Sathe, 2009) etc attempted abundance of mosquitoes from India. The present work is precise attempt on the seasonal abundance of mosquitoes and will add great relevance in solving cases of mosquito borne diseases in the region.

Materials and Methods

The survey of mosquitoes was made from Western Maharashtra (districts Kolhapur and Sangli) (Figure 2) from 2007 to 2013. A large number of specimen were collected by visiting various places of Western Maharashtra namely, districts Kolhapur (Ajra, Malkapur, Kagal, Kolhapur and Jaysingpur) and Sangli (Miraj, Vita, Tasgaon, Shirala and Jath) at 15 days interval.

The mosquito surveillance was carried out indoor as well as outdoor. Mosquito surveillance started in early in the morning from 6.15 am or in evening after 6.30 pm. Mosquitoes were collected by suction tube were transported in to test tubes for further identification. Larvae were collected with the help of ladle and dropper by one-man one-hour density. Larvae and pupae of mosquitoes from natural habitats from selected spots have also been collected and reared in the laboratory for their adult formation. The specimen collected during study period were identified by consulting Christophers (1933), Barraud (1934), Horsfall (1955), Rao (1984), Nagpal and Sharma (1994), Sathe and Girhe (2002) and Sathe and Tingare (2010) Distribution records of the specimens have been made by visiting, collecting and identifying the species from study spots of Western Maharashtra.

Results

Results are recorded in Table 1 and Figures 20–35.

The observations on seasonal abundance of mosquito species belonging to genera *Anopheles, Culex* and *Aedes* indicates that out of 31 species, 10 species were rare and 21 species were common in the Western Maharashtra.

Table 1: Mosquitoes from Southern Maharashtra.

Sl.No.	Mosquito Species	Abundance	Citations
FAMILY - CULICIDAE			
SUB FAMILY - ANOPHELINAE			
GENUS - *ANOPHELES*			
1.	*Anopheles culicifacies* Giles	Common	1901b. *Ent. Mon. Mag.* 37 : 196-198.
2.	*Anopheles stephensi* Liston	Common	1901. *Indian Med. Gat.* 36 : 361-366.
3.	*Anopheles annularis* Vander Wulp.	Common	1884. *Notes Leyden Mus.* 6 : 248-256.
4.	*Anopheles subpictus* Grassi	Common	1899. *Indian Entomologist* 34 : 192-197.
5.	*Anopheles turkhudi* Liston	Rare	1901. *Indian Med. Gat.* 36 : 441-443.
6.	*Anopheles compestris* Reid.	Common	1962. *Notes Leyden Mus.* 6 : 248-256.
7.	*Anopheles culiciformis* Cogill	Common	1903. *J. Bombay nat. Hist. Soc.* 15 : 327-336, 1 pl.
8	*Anopheles jeyporiensis* James	Rare	1902. *Sci. Mem. Med. Sanit. Dept. India (N.S.)* No. 2, 106 pp.
9.	*Anopheles karwari* James	Rare	*1902. Sci. Mem. Med. Sanit. Dept. India (N.S.)* No. 2, 106 pp.
10.	*Anopheles maculatus* Theobald,	Rare	1901. *A mon. of Culicidae or Mosq,* 1: 171-174.
11.	*Anopheles vagus* Doenitz.	Rare	*1902. Zeit. Fur Hyg. Und Infek.,* 41: 15-88.
12.	*Anopheles mahabaleshwari* sp. nov.	Common	
13.	*Anopheles waii* sp. nov.	Common	
14.	*Anopheles karveeri* sp. nov.	Common	
15.	*Anopheles krishnai* sp. nov.	Common	
16.	*Anopheles kolhapuri* sp. nov.	Common	

Contd...

Table 1–*Contd...*

Sl.No.	Mosquito Species	Abundance	Citations
SUB FAMILY - CULICINAE			
GENUS - *CULEX*			
17.	*Culex epidesmus* Theobald	Rare	1910a. *Royal Society* 12 pp. *British Museum* (*Nat. His.*).
18.	*Culex tritaeniorhynchus* Giles	Common	1901a. *J. Bombay Soc.* **13** : 592-610, pls. A and B.
19.	*Culex vishnui* Theobald	Common	1910. *Rec. Indian Mus.* **4** : 1-33, 3 pls.
20.	*Culex quinquefasciatus* Say	Common	1823. *J. Acad. Nat. Sci.* Philad **3** : 9-54.
21.	*Culex fuscocephala* Theobald	Rare	1907. M. C. iv. P. 420.
22.	*Culex malhari* sp. nov.	Common	
23.	*Culex malkapuri* sp. nov.	Common	
24.	*Culex satarensis* sp. nov.	Common	
25.	*Culex mirjensis* sp. nov.	Common	
SUB FAMILY - CULICINAE			
GENUS – *ARMIGERES*			
26.	*Armigeres* (*Armigeres*) *subalbatus* Coquillett.	Common	1898. *Rec. Indian Mus.* **4** : 1-33, 3 pls.
SUB FAMILY - CULICINAE			
GENUS - *AEDES*			
27.	*Aedes aegypti* Linnaeus	Common	1762. Zweyter Theil, ent. Besc. varschiedener wichtiger Naturalien pp. 267-606.
28.	*Aedes albopictus* Skuse	Common	1894. *Indian Mus. Notes.* **3**, No. 5, p. 20.
29.	*Aedes vittatus* Bigot	Rare	1861. *Ann. Soc. ent. Fr.* (4) **1** : 227-229.
30.	*Aedes* (*Mucidus*) *sathei* sp. nov.	Rare	
31.	*Aedes* (*Finalaya*) *rajashri* sp. nov.	Rare	

Among the total 3362 mosquito species collected the highest contribution was from *Anopheles* (45.33 per cent) followed by *Culex* (35.01 per cent), *Armigeres* (10.71 per cent), and *Aedes* (8.95 per cent). The remaining 26 species were contributes 41.7 per cent. Five vector species were among the 16 *Anopheles* species (55 per cent), three vector species (93.02 per cent) among the 5 *Aedes* species and four vector species (52.7

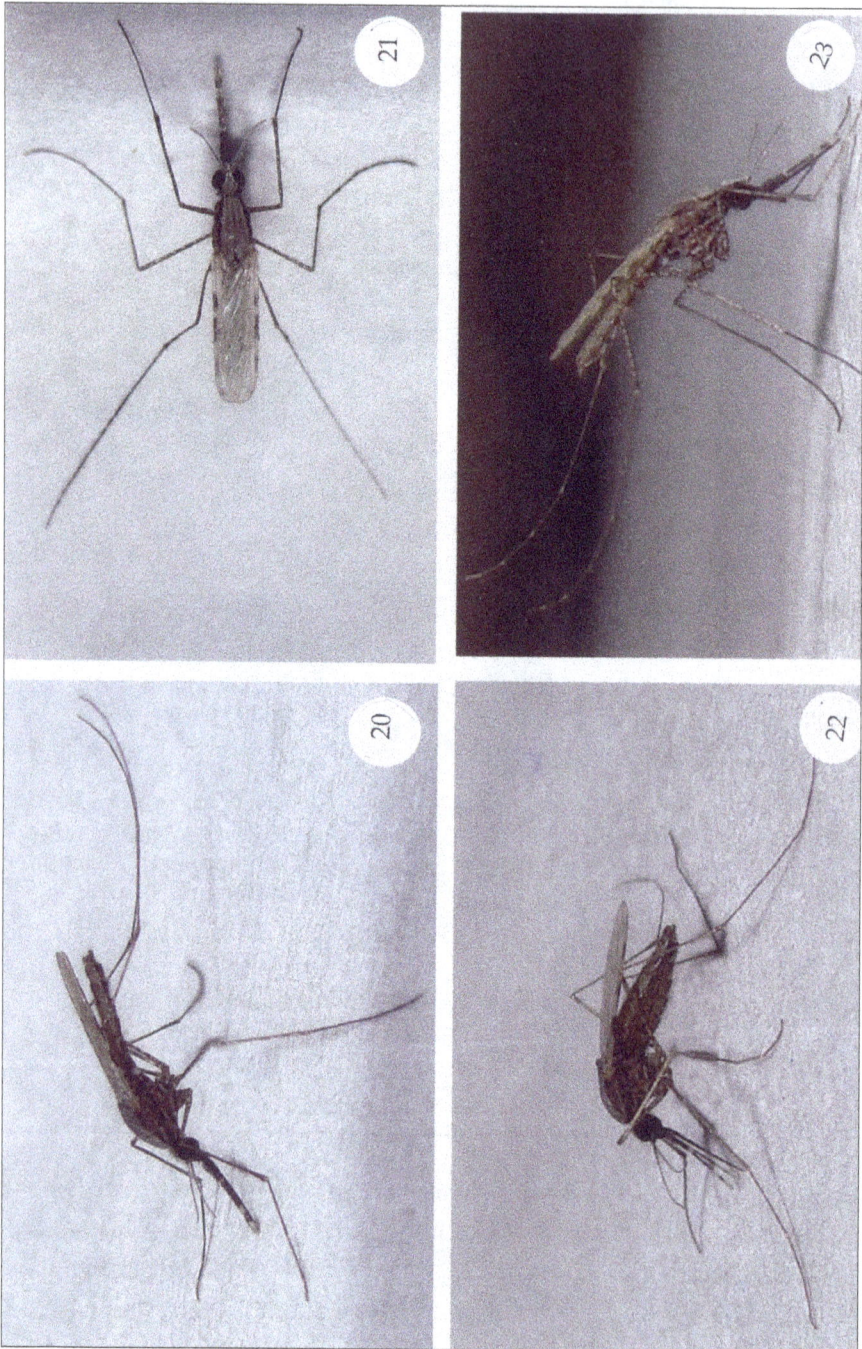

Figure 20: *Anopheles (Cellia) culicifacies*; **Figure 22**: *An. (Cellia) mahabaleshwari* sp.nov.; **Figure 23**: *An. (Cellia) waii* sp.nov.; *An. (Cellia) karveeri* sp.nov.

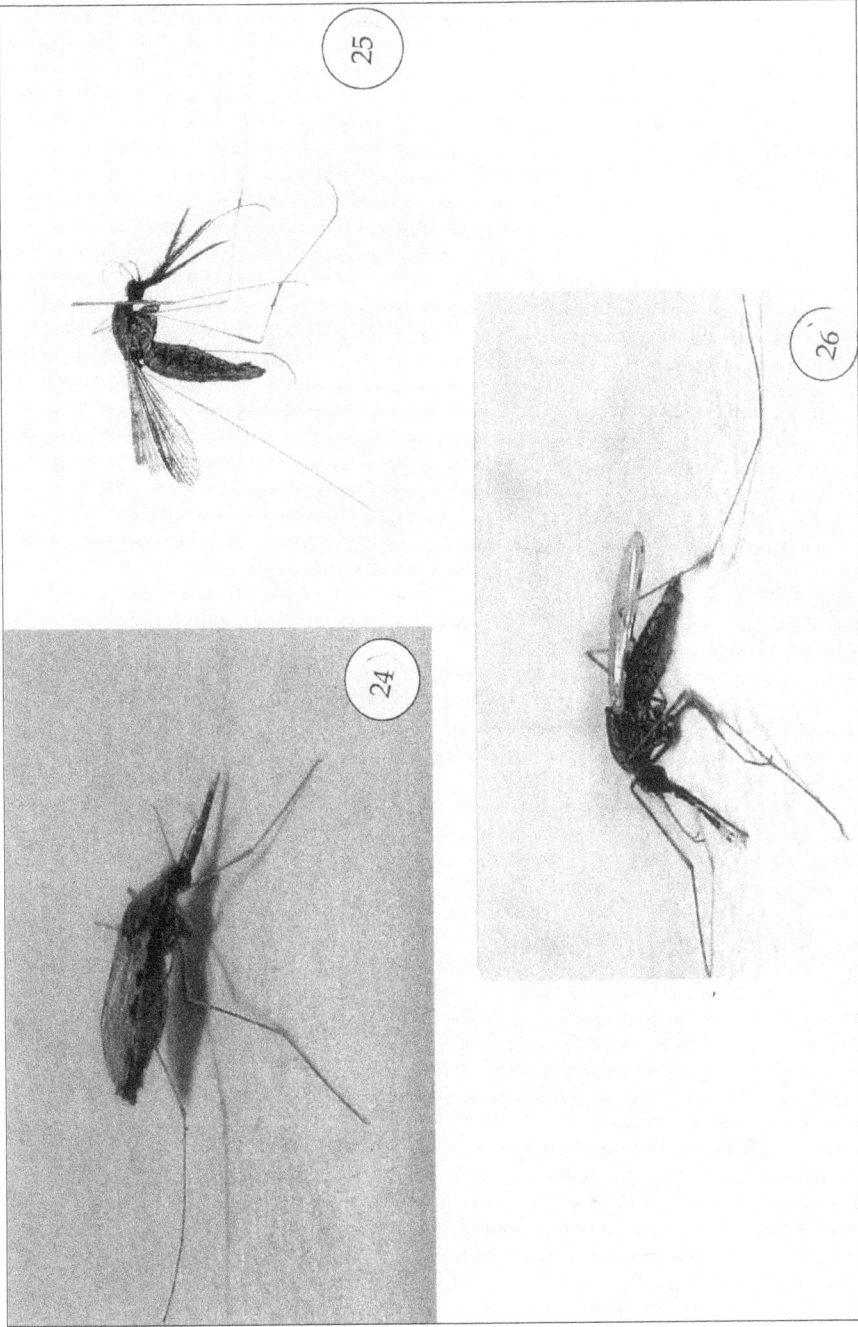

Figure 24: *Anopheles (Cellia) krishnai* sp.nov.; **Figure 25**: *An. (Anopheles) compestris;* **Figure 26**: *An. (Anopheles) kolhapuri* sp.nov.

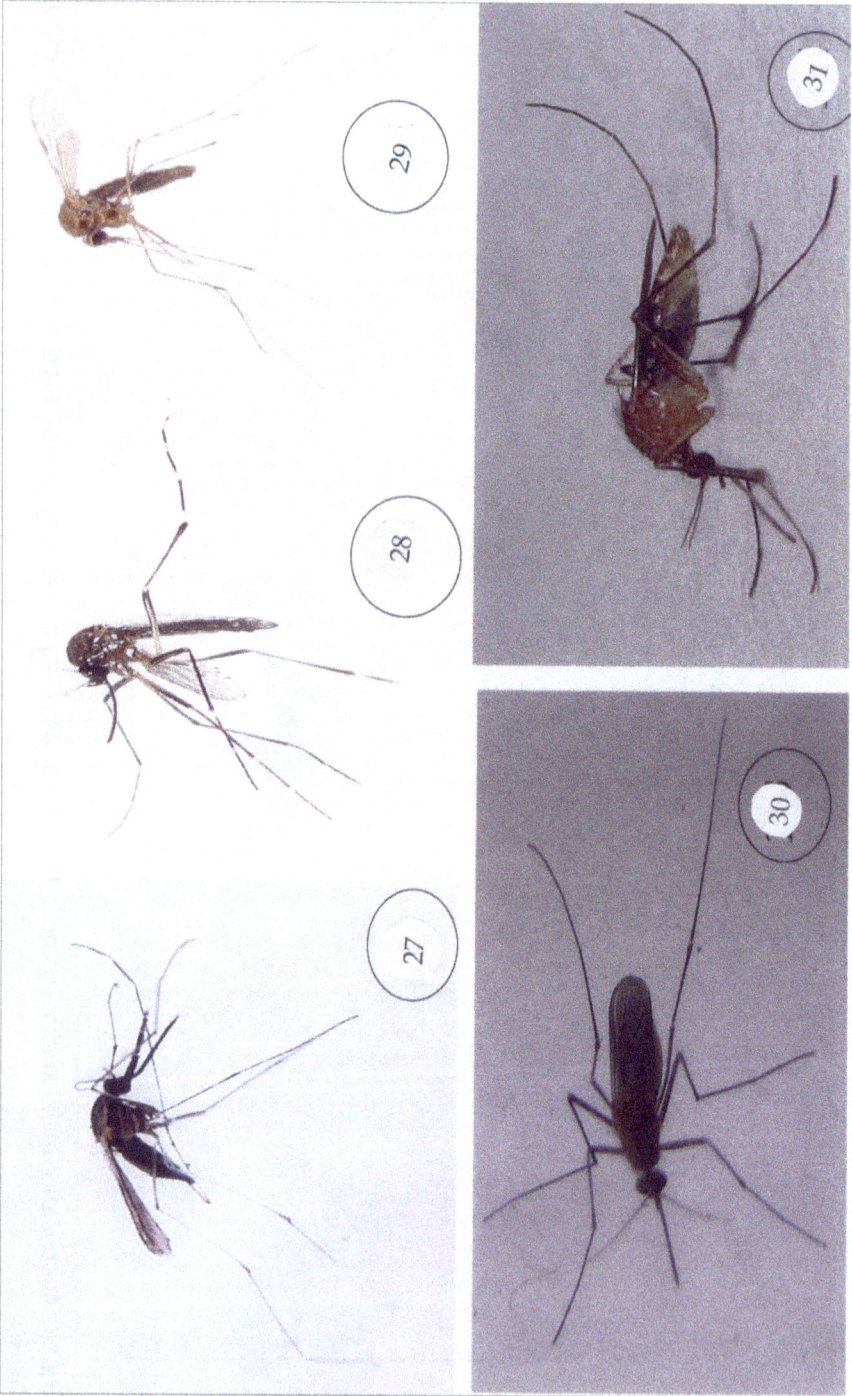

Figure 27: *Culex (Culex) quinquifasciatus*; **Figure 28**: *C. (Culex) malhari* sp.nov.; **Figure 29**: *C. (Culex) malkapuri* sp.nov.; **Figure 30**: *C. (Barraudius) mirajensis* sp.nov.; **Figure 31**: *C. (Barraudius) satarensis* sp.nov.

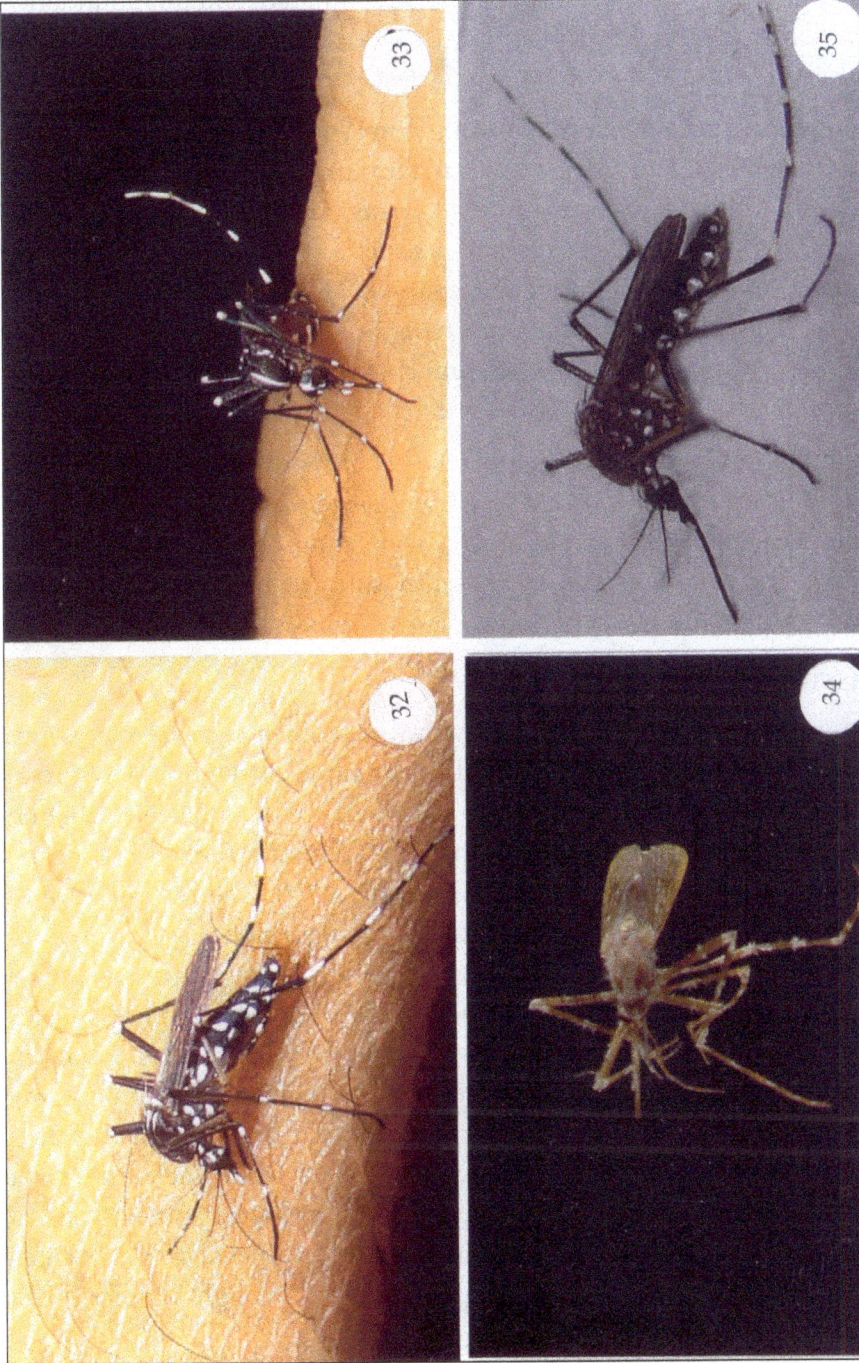

Figure 32: *Aedes (Stegomyia) aegypti*; **Figure 33**: *Ae. (Stegomyia) albopictus*; **Figure 34**: *Ae. (Mucidus) sathei* sp.nov.;
Figure 35: *Ae. (Finalagya) rajashri* sp.nov.

per cent) among the 9 *Culex* species were found in Western Maharashtra. The total 12 vector species were contributed 51.93 per cent population. Results indicate that the vector species were prominent and this abundance were alarming sign for the mosquito borne diseases in Western Maharashtra.

The seasonal prevalence of mosquitoes in Western Maharashtra reveals that the densities of *An. stephensi* was maximum during the pre monsoon period (Feb – May). The density of *An. subpictus, An. annularis, Culex vishnui, Culex tritaeniorhynchus, Culex bitaeniorhynchus* were prominent during the post monsoon period (Oct-Jan). The density of *An. culicifacies, An. fluviatilis, Aedes albopictus, Aedes aegypti, Aedes vittatus, Culex quinquefasciatus* and singular *Armigeres subalbatus* were found abundant in monsoon period (Jun – Sept).

Among newly described species *Anopheles kolhapuri, Anopheles karveeri, Anopheles krishnai, Aedes rajashri, Culex malkapuri, Culex satarensis* and *Culex mirjensis* were found abundant in monsoon period. *Anopheles mahabaleshwari, Anopheles waii, Aedes sathei* and *Culex malhari* were found abundant in post monsoon period.

Discussion

Mosquitoes respond to local temperature increase in various ways. Within limits higher temperature leads more rapid development of larval populations and shorter time between the blood meals, quicker incubation time for pathogen infection and shorter life span of adults although the later is dependent on the humidity (Roussel, 1998). The temperature of breeding places plays an important role in the persistence and growth of larvae. Rate of development of larvae accelerates in the warm water and slows down in the cold water (Ramchandra Rao, 1984). Many species such as *An. culicifacies, An. subpictus, An. vagus* breed and survive both in open water such as burrow pits or river bed pools and also in some shady places while *An. fluviatilis* and *An. minimus* prefer to breed in shady places such as under overhanging trees and bushesh or from thick growth of grass. Species breed in deep wells such as *An. stephensi* and *An. varuna* remains in shade most of the day. In fact, *An. stephensi* grows well in cistern or covered wells, which never get any direct sunlight (Ramchandra Rao, 1984).

Vector control requires through knowledge on the ecology of the local species with respect to breeding and resting habitats and behaviour. Therefore, periodic survey of vector populations in a given area is most essential for better understanding of the changing ecology, bionomics of mosquitoes and thereby possible disease outbreak can be predicted, and an effective vector control could be initiated. Keeping in view all above facts the present work was carried out. The rate of mosquito borne diseases again depends on the index of species of the region. The rapid separation and identification of mosquitoes of primary medical importance is an important task in the assessment of disease potential area.

Senior White (1937) reported twenty three anopheline species in 1935-36 on Jaypore Hills in which *Anopheles fluviatilis*, *An. varuna* had the highest peak in the month of Sept. and Feb. These species was most prevalent in the spring and in the rains. *An. jeyporiensis* and *An. culicifacies* were found resting mostly in cattle shade. Senior White *et al.* (1943) again reported eighteen anopheline species in Orissa during the year 1935-1941. *Anopheles annularis* and *An. aconitus* were most prevalent species observed in November-March.

According to Foot and Cook (1959) mosquitoes have thirty two regions in the world. In India and Sri Lanka *Anopheles culicifacies* is a most important and wide spread vector of malaria. This species was also most important vector in Srilanka. In the foot hill areas of peninsular India, *Anopheles fluviatilis* was dominant vector species for causing malaria. Foot and Cook (1959) also visualized *Anopheles varuna* from East-central India. *Anopheles stephensi* from Northern West-coast and Gangas plain and *Anopheles sundaicus* from Kolkatta, while, from Orissa they documented *Anopheles annularis*. *Culex annulifera* was the chief vector of filaria in the lower parts of the Gangas River basin, Bihar and Orissa on the North-east coast and Travancore state. Foot and Cook (1959) reported that *Aedes aegypti* was common throughout India. However, they stated that, *Aedes albopictus* was not so closer to man and less universally distributed in India.

Seasonal abundance of 11 species of *Anopheles* mosquitoes have been reported by Sen *et al.* (1960) from Dhanbad area from 1953 to 1958. *Anopheles culicifacies* was most abundant during monsoon from July to September and also in the month of February. *Anopheles subpictus* was the most predominant species and was found throughout the year with definite seasonal abundance during June to september. *A. annularis* was found during winter. *Anopheles pallidus* was found throughout the year but the peak densities were during winter (November).

Nagpal and Sharma (1983) reported seasonal abundance of twenty four species belonging to five genera of mosquitoes in the South, middle and North Andaman Island during a study tour in January-February 1982. In sixteen species of *Anopheles* the most prevalent species was *Anopheles vagus* followed by *Anopheles kochi* and *Anopheles sundaicus* and most dominated during monsoon. Among Culicines species the most dominant genera was *Culex* and most prevalent species was *Culex quinquefasciatus* followed by *Culex tritaeniorhynchus* and *Culex vishnui* and dominant during the summer and fall in the winter. The genus *Aedes, Armigers* and *Mansonia* reported extremely low in the study area.

Similarly *Anopheles vagus* and *Anopheles subpictus* exhibited a peak during the months of monsoon rains, while *An. culicifacies* a rural vector showed two peaks of abundance, one during the monsoon and other in February (Kaul *et al.,* 1982). Nagpal (1983) also studied seasonal abundance of twenty nine mosquitoes from Nainital Terai (U.P.) belonging to 8 genera *viz. Anopheles* (18), *Aedeomyia* (1), *Aedes* (2), *Armigeres* (1), *Coquillettidea* (1), *Culex* (4), *Mansonia* (1) and *Mimomyia* (1) during 1980-82. Survey revealed that *Anopheles subpictus* dominant species during Sept. 1980 and Sept-Oct. 1981, then followed by *Anopheles culicifacies* and *Anopheles fluviatilis.* However, during May-June 1981 survey *Anopheles culicifacies* was most prevalent species followed by *Anopheles subpictus* and *An. annularis.* During the Jan. – Feb. 1982 survey, *Anopheles fluviatilis* was most prevalent species followed by *Anopheles splendidus* and *Anopheles culicifacies. Culex quinquefasciatus* was the most prevalent species among the Culicinae collections in all four surveys followed by *Culex tritaeniorhynchus* and *Culex vishnui.* The seasonal abundance of *Anopheles culicifacies* and *Anopheles fluviatilis* were distributed throughout the Ferial belt, while *Anopheles culicifacies* exhibited two seasonal peaks *i.e.* in May-June and Sept. – Oct.

Das *et al.* (1984) studied seasonal abundance of forty two species belonging to a genera *viz., Anopheles, Culex, Aedes, Mansonia, Armigeres* and *Coquillettidia* of mosquitoes from various places of Meghalaya during

April-May 1980. Out of which *Anopheles vagus, Anopheles annularis, Anopheles barbirostris, Anopheles philippinensis, Culex tritaeniorhynchus, Culex vishnui, Culex bitaeniorhynchus, Culex gelidus* were most prevalent species throughout year and *Mansonia indiana, Mansonia uniformis* and *Aedes albopictus* were extremely rare.

Fifty one species of *Anopheles* and their locality, taxonomy, seasonal abundance, distribution, adult bionomics, larval ecology and diseases have been attempted by (Rao, 1984). Nagpal and Sharma (1987) documented seasonal abundance of sixty one species from Assam, Meghalaya, Arunachal Pradesh and Mizoram during Sept. 1986. These sixty one species belongs to eight genera *viz. Anopheles, Aedes, Armigeres, Coquillettidia, Culex, Malaya, Mansonia* and *Toxorhynchites*. The most dominant genus was *Anopheles* followed by *Culex, Aedes* and *Mansonia*. The most prevalent species in the genus *Anopheles* was *Anopheles vagus* followed by *Anopheles nigerrimum* and *Anopheles nivipes*. In the genus *Culex* the most prevalent species was *Culex quinquefasciatus* followed by *Culex tritaeniorhynchus* and *Culex vishnui*. *Aedes albopictus* was the most dominant species in the genus *Aedes* and it was followed by *Aedes chryolineatus* and *Aedes aegypti*. In the genus *Mansonia* the most common species was *Mansonia annulifera* followed by *Mansonia uniformis* and *Mansonia indiana*. The other four genera *viz. Armigeres, Coquillettidia, Malaya* and *Toxorhynchites* were rare in the region.

Proliferation of mosquitoes is determined by the availability of suitable and sufficient habitat for the larval stages, resting and feeding sources nearby. In urban areas like Delhi the *Aedes aegypti* populations increased with the onset of monsoon rainfall in June – July (Reuben *et al.*, 1973). Prolific breeding could be seen in man made habitats including air coolers. Ruben *et al.* (1992) visualized ten species of *Culex* mosquitoes in Madurai, Southern India. *Culex tritaeniorhynchus, Culex pseudovishnui* and *Culex vishnui* were feed dominantly on cattle, but less frequently on humans and on pigs and birds. These three species were occurring predominantly throughout year where the cattle were reared, but *Culex tritaeniorhynchus* and *Culex vishnui* showed a marked in the population during the hot season. *An. culicifacies* was recorded in high numbers from April to September while *An. fluviatilis* only during October to March in Uttaranchal (Shukla *et al.*, 2001).

From Maharashtra recently, Sathe and Girhe (2001) reported four species of *Anopheles* namely, *Anopheles culicifacies, Anopheles stephensi, Anopheles theobaldi* and *Anopheles subpictus*. The most prevalent species of *Anopheles* in Kolhapur region was *Anopheles culicifacies* while, *Anopheles subpictus* was rare in the Kolhapur region.

Sathe and Girhe (2002) reported fifteen species of mosquitoes from Kolhapur district belonging to genera *Anopheles* (4) *Culex* (3) and *Aedes* (7). Out of which *A. culicifacies, C. pipiens* and *A. aegypti* were predominant through out the year. While, *A. indica* S. and G. the largest mosquito species found was extremely rare. Other eleven species were moderately distributed in Kolhapur.

Girhe and Sathe (2001) studied incidence of malaria during the year 1992-1996, was increasing in order. Maximum, 700 infection cases were reported during the year, 1996 due to prevalence of *Anopheles* mosquitoes. Later, incidence of malaria declined from the years 1997-2000 from Kolhapur region.

Recent work of Sathe and Girhe (2001, 2002) refers to the following species of genus *Aedes* namely, *Aedes aegypti, Aedes kolhapurensis, Aedes indica, Aedes indicus, Aedes sangiti, Aedes panchgangi* and *Aedes uniformis* from Maharashtra. The work of Sathe and Girhe (2002) reported the first record of largest species *Aedes indica* from the world. *Aedes aegypti, Aedes indicus, Aedes uniformis* are well known to science from Southern Maharashtra. Four species of *Culex* namely *Culex epidesmus, Culex pipiens, Culex modestus, Culex malayi* previously have been reported from Maharashtra (Sathe and Girhe, 2002).

Murty *et al.* (2002a) studied the seasonal prevalence of *Culex quinquefasciatus* in the rural and urban areas of the East and West Godavari districts of Andhra Pradesh, India during 1999. These species occur dominantly throughout year in rural and urban areas. Murty *et al.* (2002b) reported a seasonal abundance of *Culex vishnui* sub group and *Anopheles* species in an endemic district of Andhra Pradesh during 1999. *Culex vishnui* subgroup were dominant throughout the year. Their density was high in January-December 1999. Kanojia *et al.* (2003) reported seasonal abundance of mosquito in Gorakhpur district, Uttar Pradesh during 1990 to 1996. The seasonal fluctuations in the mosquito population were recorded. High prevalence of *Culex quinquefasciatus* was observed in March, *Culex tritaeniorhynchus* predominant specieswas noticed in September. The other species such as *Culex pseudovishnui*, *Culex whitmorei*, *Culex gelidus* and *Mansonia uniformis* had also shown peak occurrence in September. Seasonal prevalence of *Anopheles* species *Anopheles subpictus* and *Anopheles peditaeniatus* showed high prevalence during July and September.

Sharma *et al.* (2005) studied seasonal prevalence of *Aedes aegypti* in Delhi during 2003. *Aedes aegypti* was abundant in month of August and September. The rise in breeding indices during the post monsoon season may be attributed to increase in artificially collected breeding containers due to rains. Joshi *et al.* (2005) gave seasonal prevalence of Anopheline mosquitoes in irrigated and non-irrigated area of Thar, Rajasthan during August 2001 to July 2002. They reported Anopheline species, *Anopheles subpictus*, *Anopheles culicifacies*, *Anopheles stephensi* and *Anopheles subpictus*. During monsoon *Anopheles subpictus* and *Anopheles stephensi* were dominant species but were peak in August to October. While *Anopheles subpictus*, *Anopheles stephensi*, *Anopheles culicifacies* and *Anopheles annularis* reported in winter season in irrigated area and *Anopheles culicifacies* and *Anopheles annularis* were predominant in summer during April to July.

Tilak *et al.* (2006) studied the seasonal prevalence of mosquito in Pune during 2001 to 2003. They reported seventeen species of five genera *Anopheles*, *Culex*, *Aedes*, *Armigeres* and *Mansonia*. The dominated genera were *Culex* followed by Anopheles, Aedes Armigeres and Mansonia. The seasonal abundance of mosquito in Pune reveals that the densities of *Anopheles stephensi*, *Anopheles varuna* and *Anopheles vagus* were maximum during summer (March – May). The density of *Anopheles annularis* and *Anopheles stephensi* were high in rainy season (June-Sept.) and winter season (Nov. – Feb.). The Culicines *Culex quinquefasciatus* was found in higher densities in all the three season with abundance in rainy season. Whereas the abundance of the other Culicines *Culex cornutus*, *Culex gelidus*, *Culex sitiens* and *Culex univittatus* were higher in summer season. *Aedes aegypti* was found in all the three seasons with high prevalence in rainy season. *Armigeres* and *Mansonia* species were extremely rare and reported in winter season.

Pemola and Jauhari (2006) reported ten *Anopheles* and *Culicine* mosquitoes in the Doon valley Dehradun, Uttarakhand during 1999-2002. The seasonal prevalence of Anopheline species *i.e. Anopheles culicifacies*,

Anopheles fluviatilis and *Anopheles stephensi* were dominant in monsoon (June-Sept.) and post-monsoon (Nov.-Dec.). The seasonal prevalence of Culicine species *Culex mimeticus, Culex vishnui, Culex quinquefasciatus* and *Aedes albopictus* were dominant in between May to November and December to February. Malarial Research Centre (MRC) (2006) reported seasonal prevalence and Bionomics of *Anopheles culicifacies, Anopheles fluviatilis, Anopheles minimus, Anopheles sundaicus* and *Anopheles stephensi* in Delhi, Kheda (Gujarat), Bhaber (Uttar Pradesh) and Rourela (Orissa) during 1989-1991. In Delhi seasonal prevalence were studied in a riverine zone of the river Yamuna and in a non riverine belt. *Anopheles culicifacies* was most dominant species in the riverine zone observed in April and in October. In non riverine area, the peak abundance was observed in May and August. *Anopheles culicifacies* was dominant in the Northern part of the reservoir zone where water pollution was at minimum level. In Khed district (Gujarat) *Anopheles culicifacies* was found throughout the year in varying proportions. In the canal-irrigated area, its density starts to build up from February and reach high in March. In the non canal-irrigated areas, the abundance of *Anopheles culicifacies* remains low throughout the year. In Bhabar area of Uttaranchal in north India, *Anopheles culicifacies* abundance remains low during January to June and October to December. It increases during monsoon reaching a high in August.

Baruah *et al.* (2007) reported seasonality of malaria in Lama camp, Hoograjuli, Behali and Pabhoi area in Sonitpur district, Assam during 2002 to 2003. The study period was grouped into four seasons such as pre monsoon (March to May), Monsoon (June to August), post Monsoon (September to November) and Winter season (December to February). They reported seven species of genus *Anopheles i.e. Anopheles annularis, Anopheles culicifacies, Anopheles dirus, Anopheles fluviatilis, Anopheles minimus, Anopheles philippinensis* and *Anopheles varuna*. The *Anopheles philippinensis* dominated in all the four study area followed by *Anopheles annularis, Anopheles minimus, Anopheles culicifacies, Anopheles fluviatilis, Anopheles dirus* and *Anopheles varuna*. Density of *Anopheles philippinensis, Anopheles annularis, Anopheles minimus, Anopheles culicifacies* and *Anopheles dirus* was increasing during the pre-monsoon period, peak in monsoon and declined during the post-monsoon.

From Southern Maharashtra Tingare and Sathe (2007) reported *Aedes khanapuri* sp.nov., *Aedes rhadhanagari* sp.nov., *Aedes tasgaonensis* sp.nov. and *Aedes mangalvedhi* sp.nov. are newly described for the first time from India. From the genus *Anopheles,* six new species have been described *i.e. Anopheles atpadi* sp.nov., *Anopheles akuluji* sp.nov., *Anopheles sagareshwari* sp.nov., *Anopheles karmalae* sp.nov., *Anopheles ajrae* sp.nov. and *Anopheles mirajensis* sp.nov. Seven species from genus *Culex* were newly described and reported from for the first time from Southern Maharashtra, which includes *Culex solapurensis* sp.nov., *Culex krishnai* sp.nov., *Culex chandrabhagi* sp. nov., *Culex rankali* sp.nov., *Culex mahalaxmi* sp.nov., *Culex kalambae* sp.nov. and *Culex sangolensis* sp.nov.

Recently, Jagtap and Sathe (2008a) studied three decades trend of malaria situation of Sangli district during the period 1971-2005. The maximum amplification of the disease was observed in drought prone area *i.e.* Jath, Kavathemahankal and Atpadi.This is correlated with rainfall. In early epidemic phase *Plasmodium vivax* was dominant but recently more than 30 per cent trend of *Plasmodium falciparum* was observed. Jagtap and Sathe (2008b) documented the role of intensified mass surveillance campaign in malaria problematic section in Sangli district. Jagtap and Sathe (2008c) studies the Chloroquine resistance to *Plasmodium falciparum* species in Etapalli block, district Gadchiroli.

Very recently, Sathe and Jagtap (2009) studied the tree hole breeding and resting of mosquitoes in Western Ghats. Total 106 tree holes were examined out of which 32 tree holes were found positive for adults and six were found for larvae. Jagtap and Sathe (2009) studied the incidence of dengue and shifting trend to rural in Kolhapur district. More recently, Sathe and Jagtap (2010) studied the abundance of Anopheles mosquitoes from Western Ghats and found responsible for malarial incidence in the region.

4

DNA and RNAs

DNA

Deoxyribose Nucleic Acid (DNA) is mostly found in nucleus, mitochondria, chloroplasts and centrioles. It is complex and largest biomolecule containing a chain of nucleotides linked to form polynucleotides. Each nucleotide is composed of pentose sugar (Deoxyribose), phosphate group and one of the four nitrogen bases attached to pentose sugar. A double helix of DNA is shown in Figure 36. DNA contains genetic information in coded form. By the process of replication DNA transfers genetic information from cell to cell. It also decides and regulates which type of proteins and protein enzymes are to be synthesized in the cell. DNA cause mutations by sudden changes in nucleotide sequence. The versatile nature of DNA is self duplication or replication. It can form the exact copies/replica of itself. During the replication of DNA following steps are noticed.

1) Replicating fork (Two DNA strands are separated)
2) Template act (Moulding for making new complimentary strand)
3) Pairing of A-T and C-G.

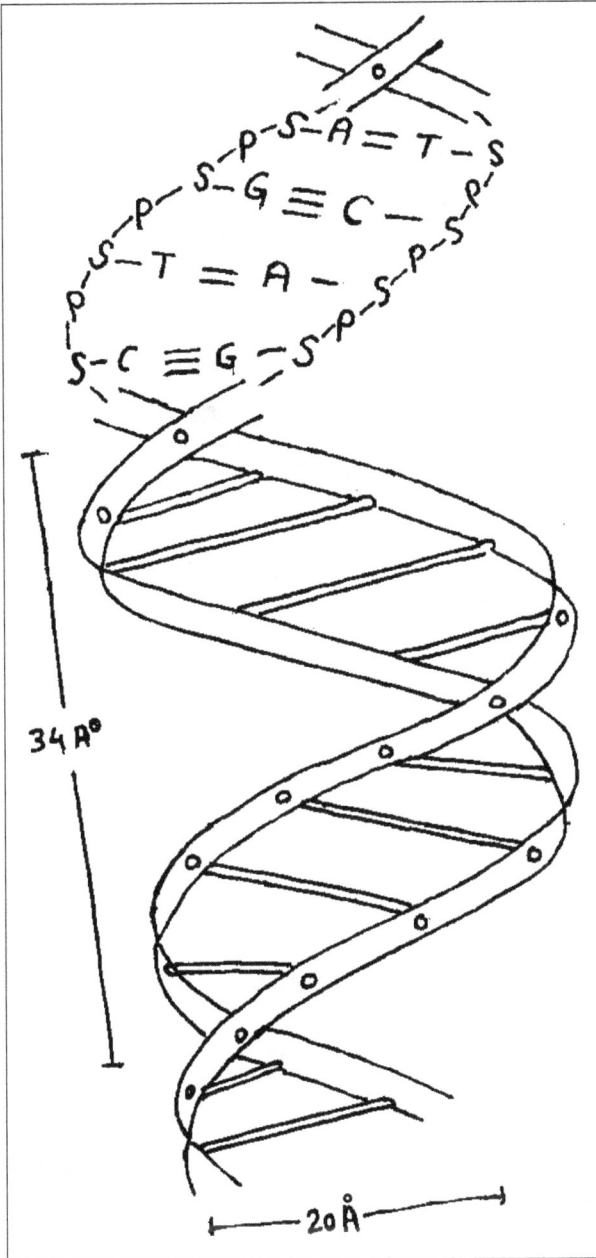

Figure 36: Double Helix of DNA.

4) Formation of part of two new or complementary strands.
5) Leading strand and lagging strand formation (Figure 37).

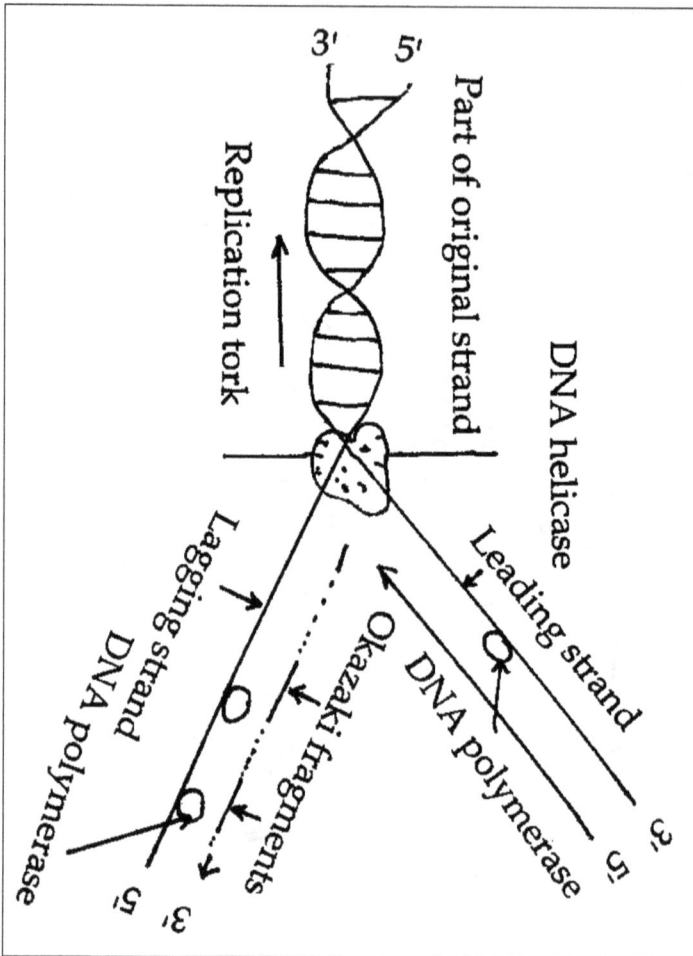

Figure 37: DNA Replication Showing Leading and Lagging Strands.

6) Continuation of replication till two new stands formed between two old ones.

7) Twisting of new and old DNA strands forming two DNA molecules.

Ribonucleic Acids (RNAs)

RNA presents in both nucleus and cytoplasm. It is single stranded, with thymine replaced by uracil and is genetic material only in some viruses. Its pentose sugar is ribose. RNA is of three types (Figure 38-40) namely,

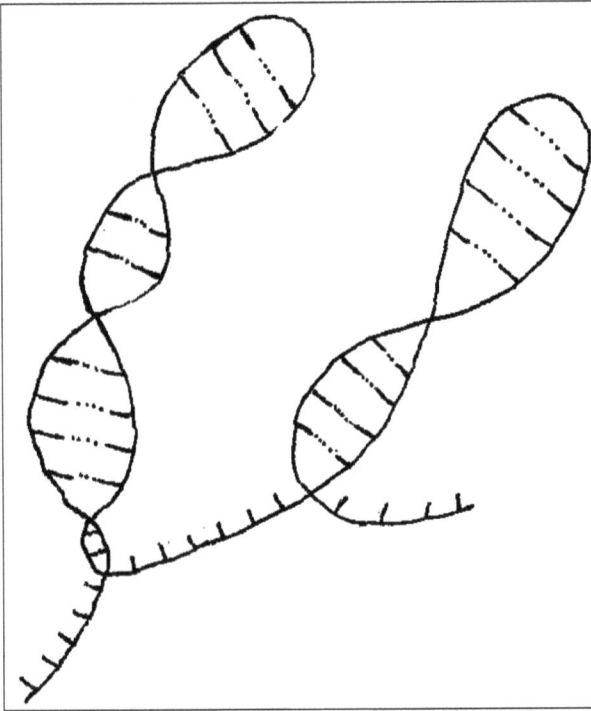

Figure 38: Ribosomal RNA (tRNA).

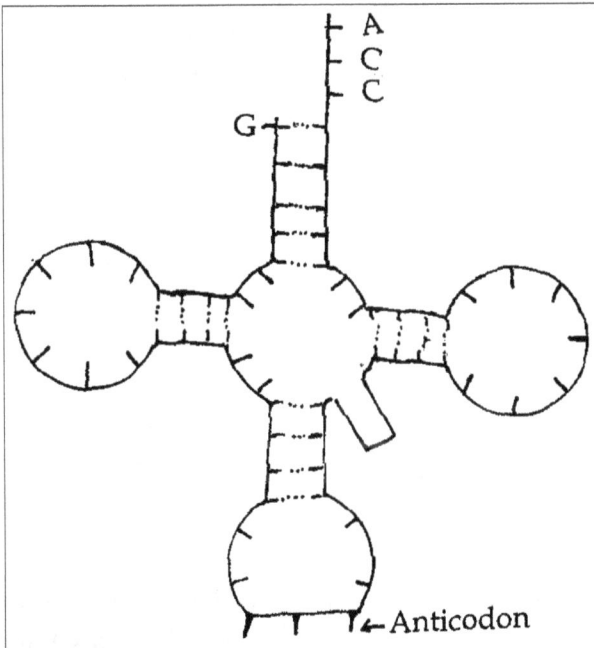

Figure 39: Transfer RNA (tRNA) (Clover leaf like).

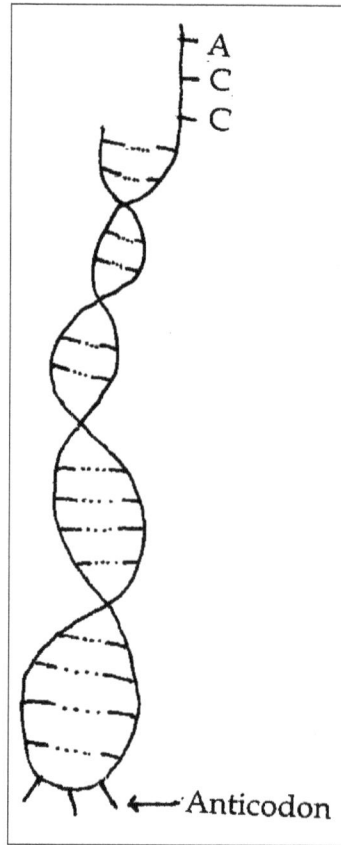

Figure 40: Transfer RNA (tRNA) (Hair pin like).

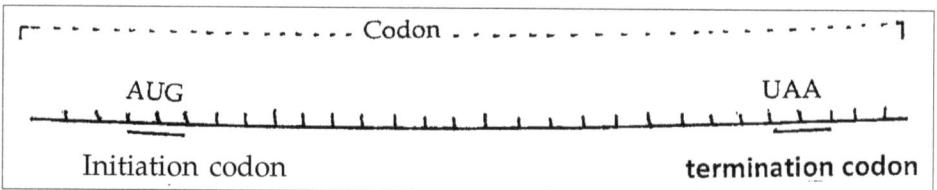

1) Messenger RNA (mRNA)
2) Ribosomal RNA (rRNA) and
3) Transfer RNA (tRNA)

Genetic Code

It is basic genetic information on DNA in code language of four alphabets A, G, C and T as nitrogen bases Adenine, Guanine, Cytosine and Thymine.

Codon

A triplet or three letter code is called Codon. A triplet code (nitrogen bases) specify one amino acid. The exact four nitrogen bases of codon are GUA, GUG, GUC and GUV for amino acid valine; UAU and UAC for tyrosine and UCG and UVG for leucine. The world codon is connected with mRNA. This nitrogen base, triplet is transcribed from DNA. Each codon codes a single specific amino acid.

Anticodon

Anticodon is related to tRNA. This nitrogen base triplet visualized as recognition site, placed on the exposed part of tRNA. On mRNA it is complementary to codon and useful component of protein synthesis.

Protein Synthesis

It takes place in two steps as following way.

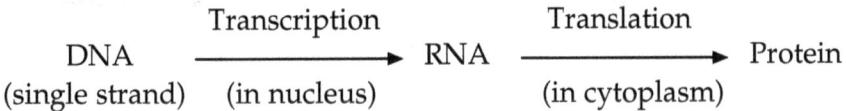

$$\underset{\substack{\text{DNA} \\ \text{(single strand)}}}{} \xrightarrow{\substack{\text{Transcription} \\ \text{(in nucleus)}}} \text{RNA} \xrightarrow{\substack{\text{Translation} \\ \text{(in cytoplasm)}}} \text{Protein}$$

According to Turner *et al.* (1998) the genetic code is the way in which the nucleotide sequence in nucleic acids specifies the amino acid sequence in proteins. It is triplet code, where the codons are adjacent and are not separated by punctuation since many of the 64 codons specify the same amino acid, the genetic code is degenerate. The genetic code is the correspondence between the sequence of the four bases of nucleic acids and the sequence of the 20 amino acids in proteins. The genetic code is generated because 18 out of 20 amino acids have been than one codon to specify them, called synonymous codons. However, methionine and tryptophan have single codons. As more gene and protein sequence information has been obtained, it has become clear that the genetic code is very nearly, but not quite, universal. This indicates that all life has evolved from single common origin. A large proportion of the DNA in eukaryotic cells does not code for protein. Most genes have large intron sequences, and genes or gene clusters are separated by large stretches of sequence of unknown function.

5

History of Molecular Evolution

The history of molecular evolution was started in the early 20th century by comparative biochemistry. However, molecular evolution field came in existence in the 1960's and 1970's with rise of molecular biology. Later, protein sequencing allowed molecular biologist to create phylogenies by sequence comparison. Differences in homogenous sequences have been used on molecular clock for estimating the time of common ancestor. A theoretical base for the molecular clock was provided by neutral theory of molecular evolution. But biologists supported mostly to panselectionism as natural selection. After 1970s nucleic acid sequencing helped molecular evolution to reach beyond proteins to highly conserved ribosomal RNA sequences. That was the foundation of reconceptualization of the early history of life.

Marcel Florkin (1940) worked on comparative biochemistry. Biochemical and quantitative data for molecular evolution was evolved in 1950s.

At the beginning of 1954, immunological assays and protein finger printing methods were developed. The primate phylogeny was studied by Williams and Goodman with this method. Later, electrophoresis and paper chromatography was used for identification of homologus proteins. Sibly applied electrophoresis for egg-white proteins in bird taxonomy. After that it was supplemented with DNA hybridization techniques as beginning of molecular systematics. Later, physical and chemical nature of genes was studied. The classical/balance controversy over the causes of heterosis, has been visualized by some workers. The effect of deleterious mutations on the average fitness of a population depends only on the rate of mutations, because, more harmful mutations are eliminated more quickly by natural selection; while less harmful mutations remain in the population for longer. This concept of genetic load was dubbed by H.J. Muller.

In 1970s and early 1980s the workers could explain the observed high levels of heterozygosity in natural population. The interaction between natural selection and genetic drift was significant for mutations. In 1970 molecular biologist supported the theory that most mutation events at the molecular level are slightly deleterious rather than strictly neutral. The molecular biologist were also in opinion that rates of protein evolution were fairly independent with generation time, rates of non coding DNA divergence were inversely proportional to generation time. Ohta proposed that most amino acid substitutions were slightly deleterious while non coding DNA substitutions were more neutral. In large populations with short generation times, non coding DNA evolved faster while protein evolution was retarded by selection.

From 1980 to 1990s 'shift model' was used in molecular evolution. In 1990's Ohta developed 'mixed model' including both beneficial and deleterious mutations, so that no artificial "shift" of overall population fitness was necessary. However, this theory was flourished after the advent of "rapid DNA sequencing". This has helped more detail studies of systematics for comparing the evolution of genome regions.

Some workers in 1960s were pushing the base of the tree of life by studying highly conserved nucleic acid sequences. Genetic code and its origin was studied by Carl Woese. He used small sub unit ribosomal RNA to classify bacteria by genetic similarity.

In 1977, it was came to know that Methanogens (Bacteria) lacks the rRNA units which was the important character for phylogeny. However, even after controversial views the work of Woese became the basis of modern three domain system of *Archaea, Bacteria* and *Eukarya*.

The importance of RNA in Microbial phylogeny suggested RNA based life had proceeded the current forms of DNA based life. Genomics produced phylogenies in 1990s contradicting the rRNA based results, recognizing lateral gene transfer across distinct taxa. However, it seems complex picture of origin and early history of life. According to Odhiambo (1987) the use of recombinant DNA provided the basis for genetic engineering, has opened up two areas of profound potential for biotechnology as applied to the life sciences.

Methodologies for the rapid analysis of complex biological molecules were developed first in 1970s. This led in 1978 to the compilation of the first computer protein atlas comprising more than 500 proteins. Secondly, the rapid synthesis of genes was opened up by the discovery of three kinds of enzymes i) enzymes capable of slicing the DNA molecule at precise sites ii) enzymes capable of sealing the loose ends of DNA fragments, and iii) enzymes having the function of synthesizing DNA from messenger RNA. These discoveries led to the rapid synthesis of numerous genes. A computer gene atlas of 350,000 genes had been compiled at the end of 1980. Following these two major molecular biological advances, it became possible to synthesize biological molecules to order by implanting specific genes into various kinds of microorganisms. Molecular biotechnology helps in several ways for control of unwanted insects and pests. Germline transformation could be used to introduce a conditional lethal gene under the control of a female - specific promoter, such that females would be eliminated from the transformed population by rearing the insects on a selective medium. X-rays induce sterility and also impair the ability of the sterile males to copulate. An anti-sense RNA approach can also play important role in pest control. The identification and characterization of transposable elements in target insects could lead to the development of specific transformation system. Sex ratio distortions and genetic inversions are possible in certain species of *Anopheles* mosquitoes. Cloning and hydridization through biotechnology can solve many problems in medical and agricultural sciences.

DNA barcoding helps in correct identification with less man power and in short time. Therefore, in practice, DNA based identification is already well established in the literature and widely accepted by research workers in biodiversity and applied sciences at global scenario.

6

DNA Sequencing/Barcoding

There was no method available for directly DNA sequencing prior to the mid 1970's. Information about gene and genome organization was based upon studies of prokaryotic organisms. The reverse genetics was the primary means of obtaining DNA sequence in which the amino acid sequence of the gene product of interest was back translated into a nucleotide sequence based upon the appropriate codons. In mid 1970's two methods were developed for directly sequencing DNA. These methods are familiar with Maxam Gilbert chemical method and Sanger Chain - termination method.

1) Maxam-Gilbert Method

Maxam-Gilbert Method was developed for sequencing single-stranded DNA under which two step catalytic processes were involved, piperidine and two chemicals that selectively attacked purines and pyrimidines. Purines reacted with dimethyl sulphate and pyrimides with hydrazine in such a way as to break the glycoside bond between the ribose sugar and the base displacing the base. Dimethyl sulfate and piperidine alone selectively cleave guanine nuoleotides but in formic acid clave both guanine and adenine nucleotides. Similarly, hydrazine and piperidine clave both thymine and cytosine nucleotides while hydrazine and piperidine in 1.5M NaCl clave cytosine nucleotides. These selective reactions for DNA sequencing helped creating a single-stranded DNA substrate carrying a radioactive label on the 5' end. This labeled substrate subjected to four separate cleavage reactions, each of which created a population labeled cleavage products ending in known nucleotides. For details Maxam and Gilbert (1977) be consulted.

2) Sanger Chain Termination Method

Fred Sanger (1977) developed chain termination method for DNA sequencing. Rather than using chemical cleavage reactions, he used a third form of the ribose sugars in the method. Ribose has a hydroxyl group on both the 2' and 3' carbons whereas deoxyribose has only the one hydroxygroup on the 3' carbon. This is not a concern for polynucleotide synthesis in vivo since the coupling occurs through the 3' carbon in both RNA and DNA. There is a third form of ribose in which the hydroxyl group is missing from both the 2' and 3' carbons. This is dideoxyribose, whenever a dideoxynucleotide was incorporated into a polynucleotide, the chain irreversibly stop or terminated. Thus incorporating specific dideoxynucleotides in vivo resulted in selective chain termination. For detail methodology Sanger (1977) be consulted.

3) DNA Sequencing Using Sequence Version 2.0 T 7

DNA Polymerase

It is done with following steps :

   i) Annealing template and primer

  ii) Labeling step

 iii) Termination step

 iv) Sequencing plasmid DNA

  v) Running sequencing reactions in 96 - well microassay plates

 vi) Improving band intensities close to the primer

 vii) Extending sequences beyond 400 bases from the primer.

viii) Compressions

 ix) For details sunsult Delidow *et al.* (Methods in Mol. Biol. 58, 275-387).

4) Automated Fluorescence Sequencing

This is most dramatic and advanced method of DNA sequencing developed by using fluorescence - labeled dideoxy-terminators. A DNA sequencing method was developed by Hood *et al.* (1986) in which the radioactive labels, autoradiography and manual base calling were all replaced by fluorescent labels, laser induced fluorescence detection and computerized base calling. In this method, the primer was labeled with one of four different fluorescent dyes and each was placed in a separate sequencing reaction with one of four different fluorescent dyes and each was placed in a separate sequencing reaction with one of the four dideoxynucleotides plus all four deoxynucleotides. After completion of reactions, the four reactions were pooled and run together in one lane of a polyacrylamide sequencing gel. A four coloured laser induced fluorescence detector scanned the gel as the reaction fragments migrated past. The influorescence signature of each fragment was then sent to a computer for base calling. However, the above method was commercialized in 1987. Later, Prober *et al.* (1987) advanced this technique by labeled terminators instead of fluorescence primers. In 1990's the technique was refined with terminator chemistries and the detection/ base calling. Changing the due labels on the terminators for improvement of fragment resolution was the important aspect.

Swerdlow *et al.* (1990) reported about the use of capillaries to obtain DNA sequences. A capillary based system can be run with much higher voltages for dramatically lowering the run times, the capillary systems can be automated, a major limitation in gel based systems. Later, in 1993 Karger *et al.*, accounted on the use of viscosity separation matrix that could be pumped into capillaries at relatively low pressure. The matrix replace cross-linked polyacryl-amide and remove the final obstacle to the development of a truly automated DNA sequencing platform. The low viscosity non-cross-linked polymer flushed out after a run without touching capillary. According to Zhang *et al.* (1995) non cross-linked polymer could be stable at 60°C and can deliver high quality sequence data. DNA sequencing reactions be carried out in a single reaction tube and prepared for loading once the reaction reagents had been filtered out.

1) The capillary be set up to deliver new polymer to the capillary.

2) Load the sequencing reaction into the capillary.

3) Apply a constant electrical current through capillary.

4) Collect the resolved fragments migrate past an optical window where a laser would excite the dye terminator.

5) A detector can collect the fluorescence emission wavelengths.

6) Software can interpret the emission wavelengths as nucleotides.

7) 500-1000 bases of high quality DNA sequence will be collected within few hours.

5) A Basic Polymerase Chain Reaction Protocol

A dozens of variations in the basic theme of PCR have successfully been carried out. However, there are many ways to do a PCR reaction. The basic, straight - forward PCR protocols are given below :

☆ Step - 1: Choosing target substrates and PCR primers.

☆ Step - 2: Setting up the reaction.

☆ Step - 3: Choosing the reaction conditions.

☆ Step - 4: Validating the reaction.

6) Neighbour - Joining Method for Phylogenic Trees

Saitou and Nei (1987) developed a new method for reconstructing phylogenetic trees from evolutionary distance data. Pairs of operational taxonomic units (OTUS = neighbours) that minimize the total branchs length at each stage of clustering of OTUS starting with a star like tree.

The branch lengths as well as the topology of a parsimonious tree can quickly be obtained by using this method. Using computer simulation they studied the efficiency of this method in obtaining the correct unrooted tree in comparison with that of five other three making methods. According to them the new neighbour-joining method and Sattath and Tversky's method are generally better than other methods.

7

Molecular Phylogeny of Mosquitoes

Introduction

Description and study of biodiversity is an important aspect of intricate evolutionary trends of life. Linnaean classifcation of animals and plants is an important step toward this goal. According to Besansky *et al.*, 2003, and Pennisi 2003 only 10 per cent of extant species on earth (10-15 million) are known to the science. As regards to the mosquitoes 3500 species of mosquitoes have been described from the world (Sathe and Tingare, 2010). Except for a few of medically important species, mosquito taxonomy has stagnated for 100 years at the stage of recognizing and naming species. Recent studies on DNA-based approaches show a promising trend in the rapid description of biodiversity (Hebert *et al.*, 2003a; Hebert and Gregory 2005). Traditional morphology-based assessments are time-consuming and require specialists whose numbers are insufficient and dwindling. A DNA-based method called DNA barcoding has been proposed as a rapid means of cataloguing species. Hebert *et al.* specifically suggest the employment of DNA sequences as taxon 'barcodes' and propose that the mitochondrial gene cytochrome oxidase-I (CO-I) serves as the core of a global bioidentification system for animals.

Barcoding is not an end in itself but it will boost the rate of discovery. The unique contribution of DNA barcoding to mosquito taxonomy and systematic is a compressed timeline for the exploration and analysis of biodiversity. According to Knowlton and Weigt (1998) among the mitochondrial genes, cytochrome *c* oxidase subunit 1 (CO-I) is to be the most conserved gene in the amino acid sequences and hence has distinct advantage for taxonomic studies. The morphological identification keys used currently for identification of mosquitoes are mainly related to imaginal and fourth instars only. Hence, difficult to identify other stages of development collected in the feld, if not reared in the laboratory. In many adult specimen bristles and hairs get damaged which make identification more difficult. The sibling species may also lead to incorrect identification of mosquitoes since they are morphologically indistinguishable and identified only by cyto taxonomic features, using polytene chromosomes, specific to certain tissues in particular developmental stages. Therefore, construction of DNA barcodes for each species of mosquito would provide an important tool for identification of mosquito species and may enable description of the species biodiversity. Examination of polytene chromosomes for species-specific diagnostic inversion genotypes is the cheapest technique now available, but it requires specialized skill and aptitude for microscopic examination.

According to Richardson *et al.* (1986) the development of DNA based marker system, has great importance in barcoding greater level of polymorphism could be obtained by using appropriate DNA markers than by using protein markers in many situation. Moreover, DNA samples are more stable than proteins and are unchanged for detection at all time and tissue of the organism unlike proteins. Thus, DNA markers became the most common yardstick for measuring genetic differences between individuals or within and between related species or populations. A singularly important aspect of molecular analysis is that they allow direct comparisons of genetic differences among any group of organisms.

Molecular genetic studies based on the polymerase chain reaction (PCR) have increasingly importance. The main molecular genetic marker

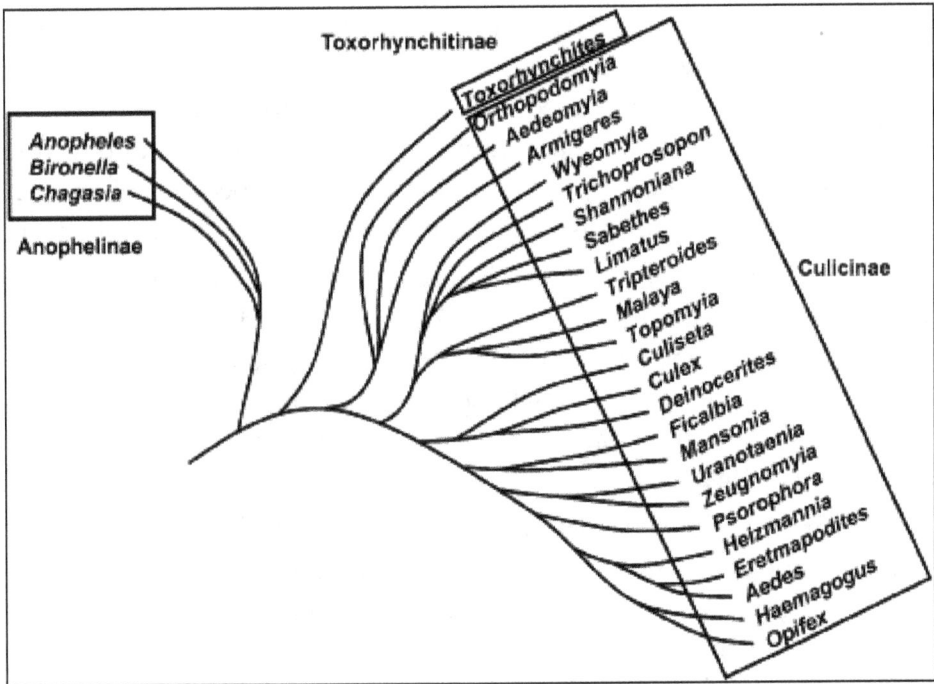

Figure 41: Intuitive Phylogeny of Mosquito Genera by Ross, 1951.

for species differentiation is the second internal transcribed spacer of the ribosomal gene cluster (ITS2). The species identification is based on an analysis of this ribosomal deoxyribonucleic acid (DNA) region by amplification with subsequent analysis of the restriction fragment length polymorphism (RFLP) and, if necessary, by sequencing the amplification products.

Ross (1951) constructed the first evolutionary tree of Culicidae based on an intuitive interpretation of comparative bionomics and morphological data (Figure 41).

The first molecular systematics research was based on immunological assays and protein "fingerprinting" methods. Edwards (1923) surmised that "The origin and phylogenetic history of the Culicidae must go back to well into the Mesozoic Era". The first evolutionary tree of Culicidae was constructed by Ross (1951). Poinar *et al.* (2000) provided a list and critical evaluation of mosquito fossils. Harbach and Kitching (1998) examined the generic and cladistic analyses of morphological data of 38 genera by analysiing of 73 characters from larvae, pupae and adults. Miller *et al*. (1997) studied the phylogenetic relationships of four mosquito species. Besansky and Fahey (1997) used the nuclear protein-coding *white* gene to examine the phylogenetic relationships of 13 species of mosquitoes. Isoe (2000) examined the phylogenetic relationships of 39 species. Sallum *et al*. (2000) performed the first phylogenetic analysis of subfamily Anophelinae.

Aedini is the largest tribe in family Culicidae with 1,240 currently recognized species. Anderson *et al*. (2001) examined the phylogenetic relationships of six mosquito species belonging to genera *Aedes* (3 species), *Armigeres* (1 species) and *Culex* (2 species) by comparing chromosomal rearrangements (Anderson *et al.*, 2001, Besansky and Fahey, 1997 and Wesson *et al.*, 1992).

Kumar *et al*. (1998), Coock *et al*. (2005), Shepard *et al*. (2006), Wesson *et al*. (1992) and Besansky and Fahey, (1997), Reinert *et al*. (2004, 2006), Zavortink, (1972), Rey *et al*. (2001) etc. worked on the phylogenetic relationships of mosquito species.

Culex is by far the largest genus with 763 species divided between 23 subgenera. Miller *et al*. (1996) were the first investigators to examine relationships among *Culex* mosquitoes based on molecular data. Belkin, (1962) and Bram, (1968), Navarro and Liria, (2000), Juthayothin, (2004), St John, (2007), etc. also worked on *Culex* molecular phylogeny.

Materials and Methods

Mosquito specimen used for constructing DNA barcodes were from collections made from different study spots (Figure 3). The adults were identified morphologically and used for DNA extraction. Species identification was done based on taxonomic keys (Christophers, 1933; Barraud, 1934; Rao, 1984).

Genomic DNA extraction was carried out from whole mosquito using the QIAamp DNA Mini Kit (QIAGEN) following the manufacturer's instructions. The Insect COI gene was PCR amplified from the total chromosomal DNA using universal primer LCO (5'- GGTCAACAAAT CATAAAGATATTGG-3') and HCO (5' TAAACTTCAGGGTGACC AAAAAATCA – 3'). The reaction mixture (25µl) was composed of 10X buffer 2.5 µl, 2 mM dNTP 2.5 µl,10 pMol/L16F271.25 µl, 10 pMol

17R1525XP 1.25 µl, 10U Taq DNA Polymerase 0.2 µl, template DNA 2L and water 15.3 L. Reactions were amplified through 35 cycles at following parameters : one minute at 95°C, one, minute at 40°C, and one and half minutes at 72°C, followed by a final extention step at 72°C, for seven minutes and PCR cycler was applied biosystem PCR system. The PCR product was purified with 0.6 volumes of PEG- NaCl and incubated for 30 min at 37°C. The precipitate was collected by centrifugation for 30 min at 12000 rpm. The pellet was washed twice with 70 per cent ethanol, dried under vacuum, resuspended in distilled water concentration of >0.1 pmol and the purified product was sequenced using BIGDYE terminator kit 3.1CABI Perkin Elmer USA. The sequencing reactions were run on ABI-PRISM 31D automated sequencer (Model 3730XL; Applied Biosystem, USA).

The 16S rRNA sequences obtained above were aligned by using BLAST analysis at NCBI server (http ://www.ncbi.nlm.nih.gov/BLAST), also BOLD database was used for analysis purpose.

The amplified fragments were run on a 1 per cent agarose gel to check the integrity of the fragments and the PCR product was purified by QIAGEN Gmb HPCR purification kit. The DNA sequences were subjected to alignment using Clustal W. Sequence divergences among individuals were quantified by using the Kimura two-parameter distance model (Kimura, 1980). A neighbourhood joining (NJ) tree of K2P distances was created to provide a graphic representation of the clustering pattern among different species (Saitou and Nei 1987, Hajibabaei *et al.*, 2006). These analyses of the sequences were conducted using MEGA version 4 software (Kumar *et al.*, 2004). The holotype and paratypes are time being with Department of Zoology and will be deposited at ZSI, Kolkatta in due course time.

Results

The neighbour-joining analysis of nucleotide and amino acid sequences of 8 mosquito species showed that most mosquito species separated into distinct clusters (Figure 42 to 49). Hence, they are supposed to be the region specific species. Their evolutionary relationship through phyllogenetic trees are presented in next pages:

1. *Anopheles (Cellia) culicifacies* Giles, 1901

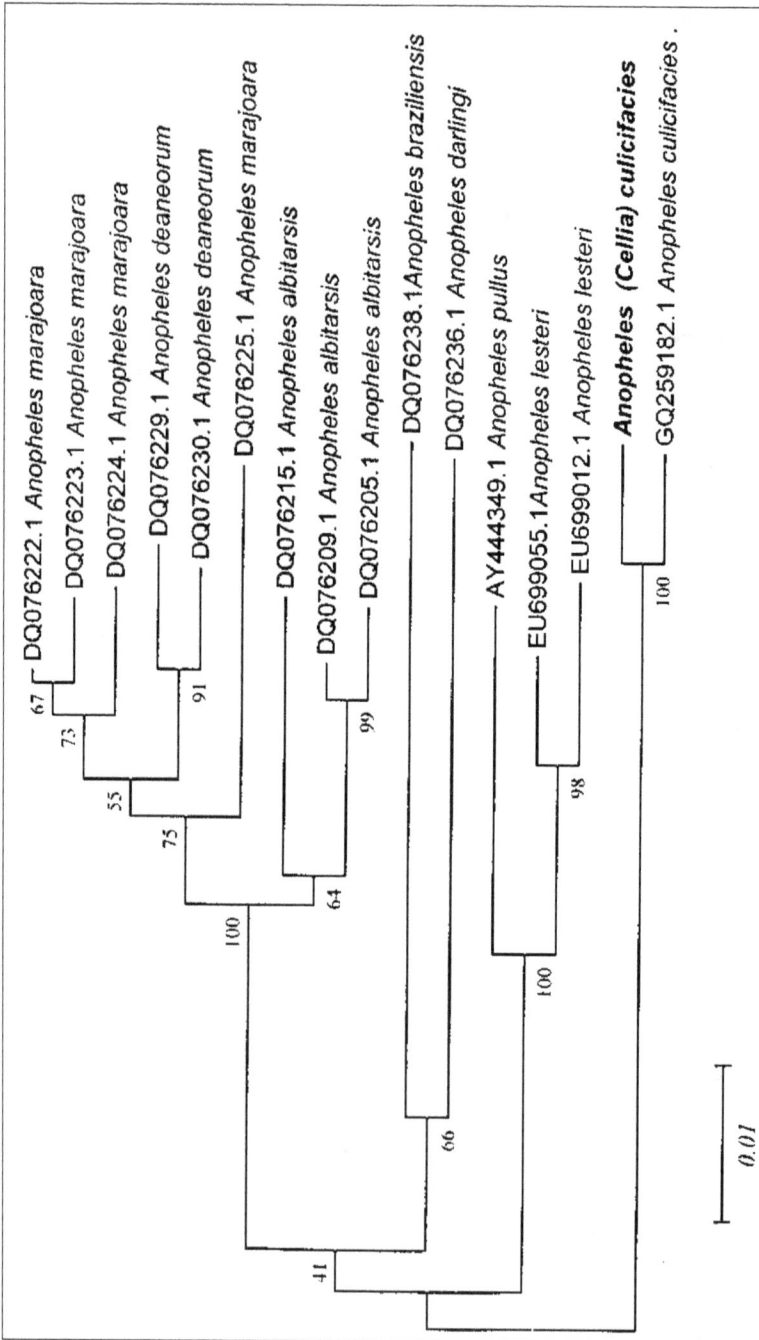

Figure 42: Evolutionary relationships of 16 taxa constructed phylogenetic tree by using neighbour joining method and MEGA 4 software.

The optimal tree with the sum of branch length = 0.27758278 is shown. There were a total of 483 positions in the final dataset.

2. *Anopheles (Cellia) krishnai sp. nov.*

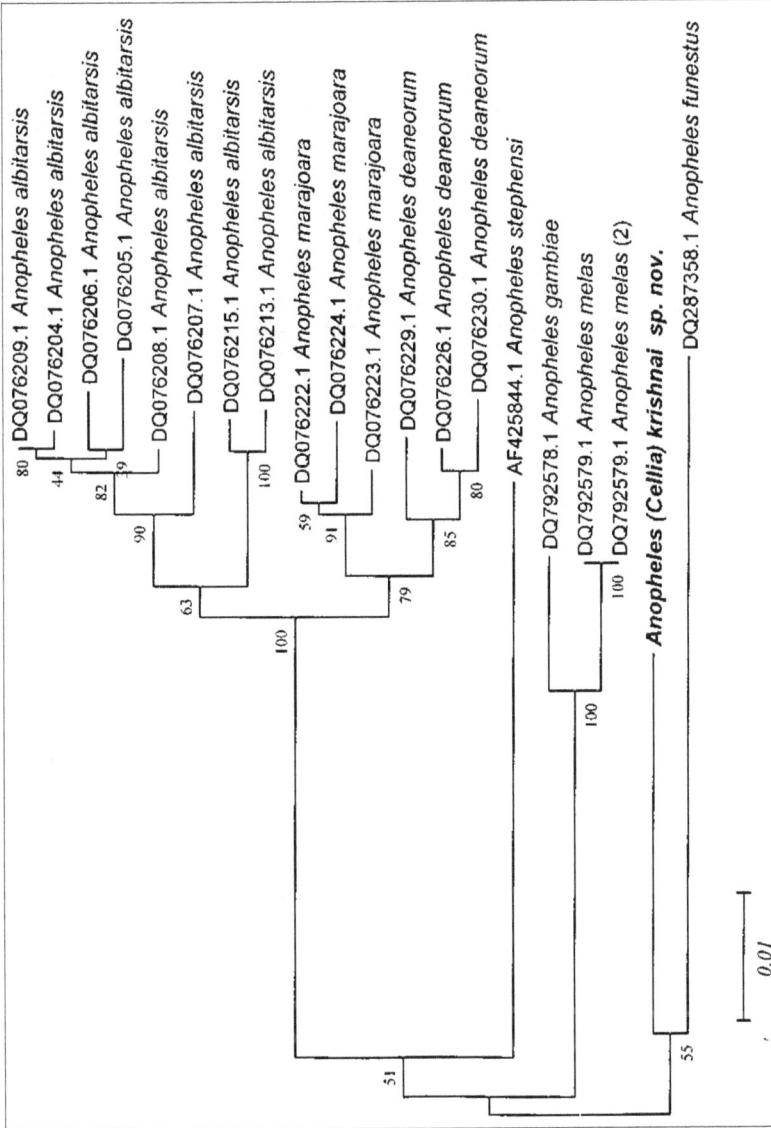

Figure 43: Evolutionary relationships of 19 taxa constructed phylogenetic tree by using neighbour joining method and MEGA 4 software.

The optimal tree with the sum of branch length = 0.29516421 is shown. There were a total of 643 positions in the final dataset.

3. *Anopheles (Anopheles) compestris* **Reid 1962.**

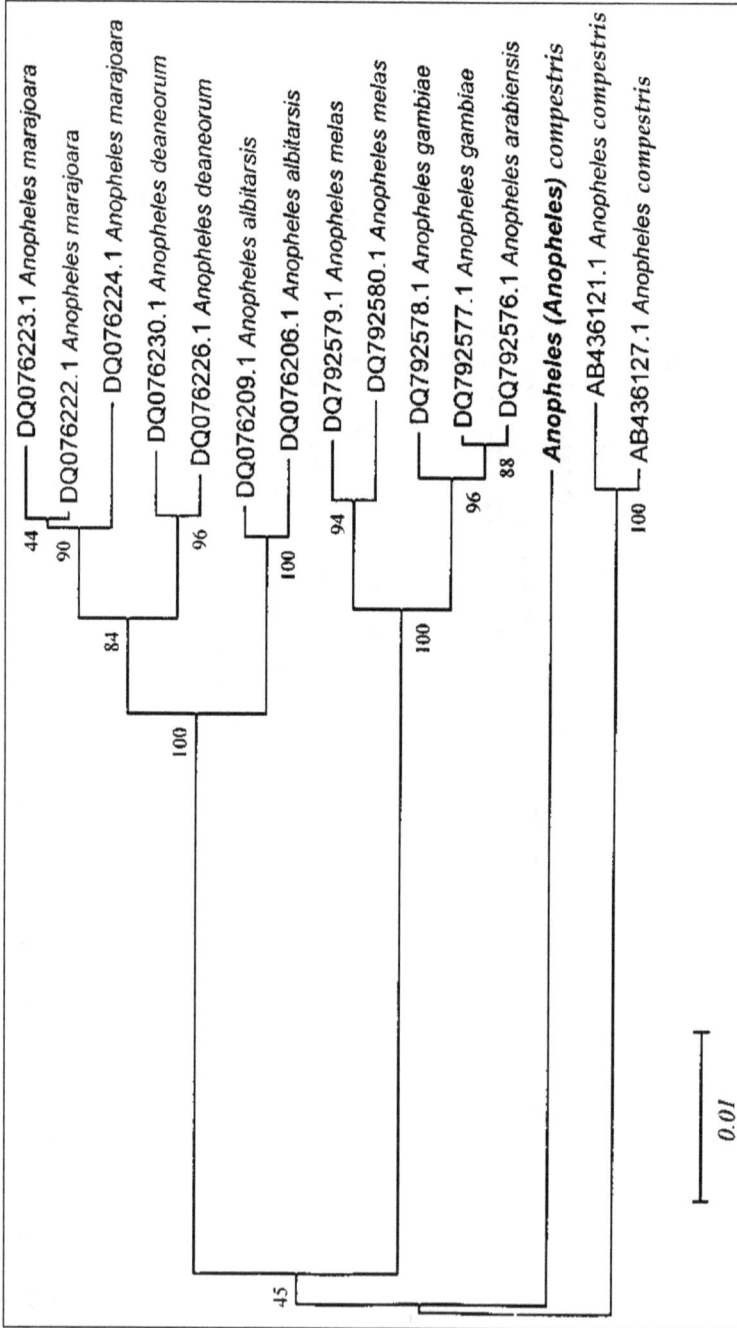

Figure 44: Evolutionary relationships of 15 taxa constructed phyllogenetic tree by using neighbour joining method and MEGA 4 software.

The optimal tree with the sum of branch length = 0.14678612 was noticed. There were a total of 631 positions in the final dataset.

4. *Aedes (Stegomyia) aegypti* Linnaeus, 1762.

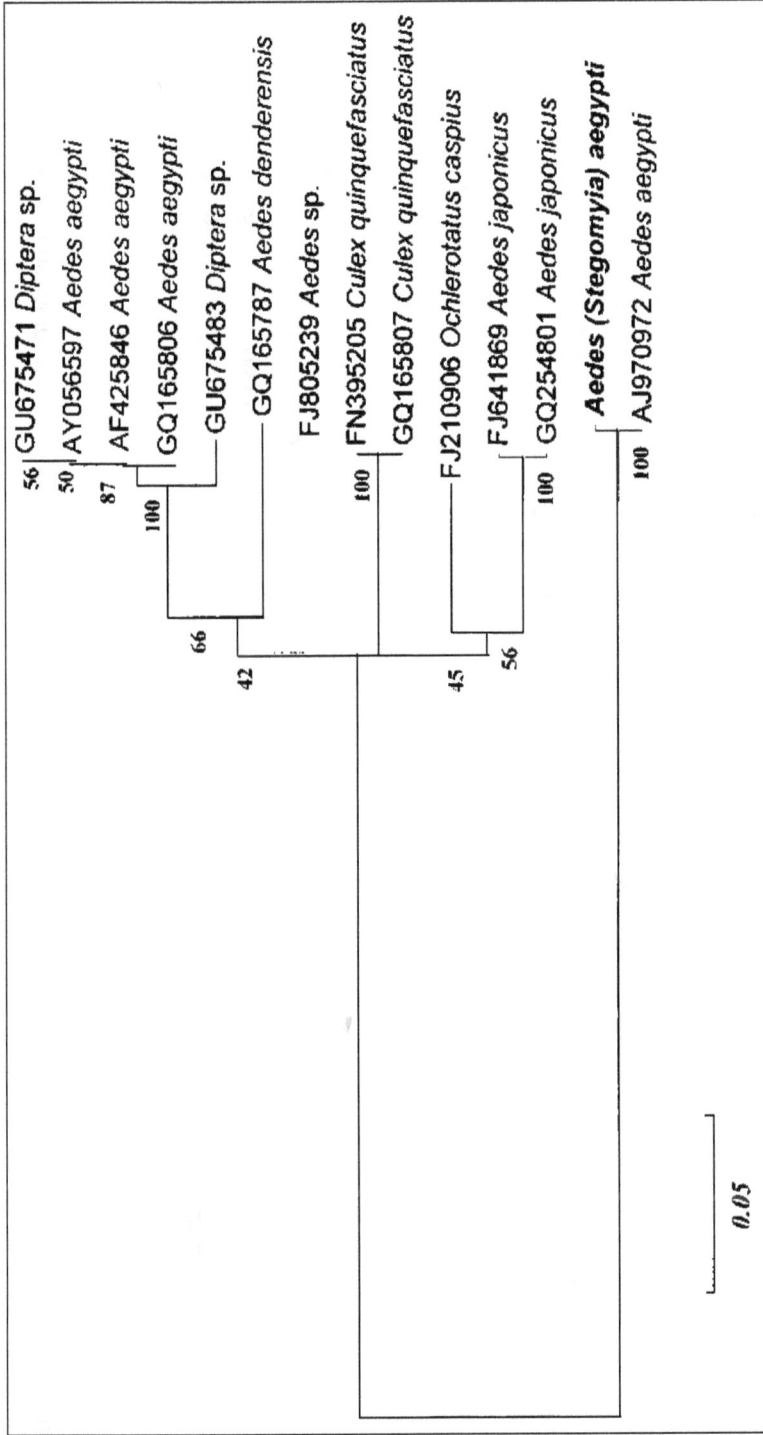

Figure 45: Evolutionary relationships of 14 taxa constructed phyllogenetic tree by using neighbour joining method and MEGA 4 software.

The optimal tree with the sum of branch length = 7.35313831 is shown. There were a total of 551 positions in the final dataset.

5. *Aedes (Stegomyia) albopictus* Skuse, 1894.

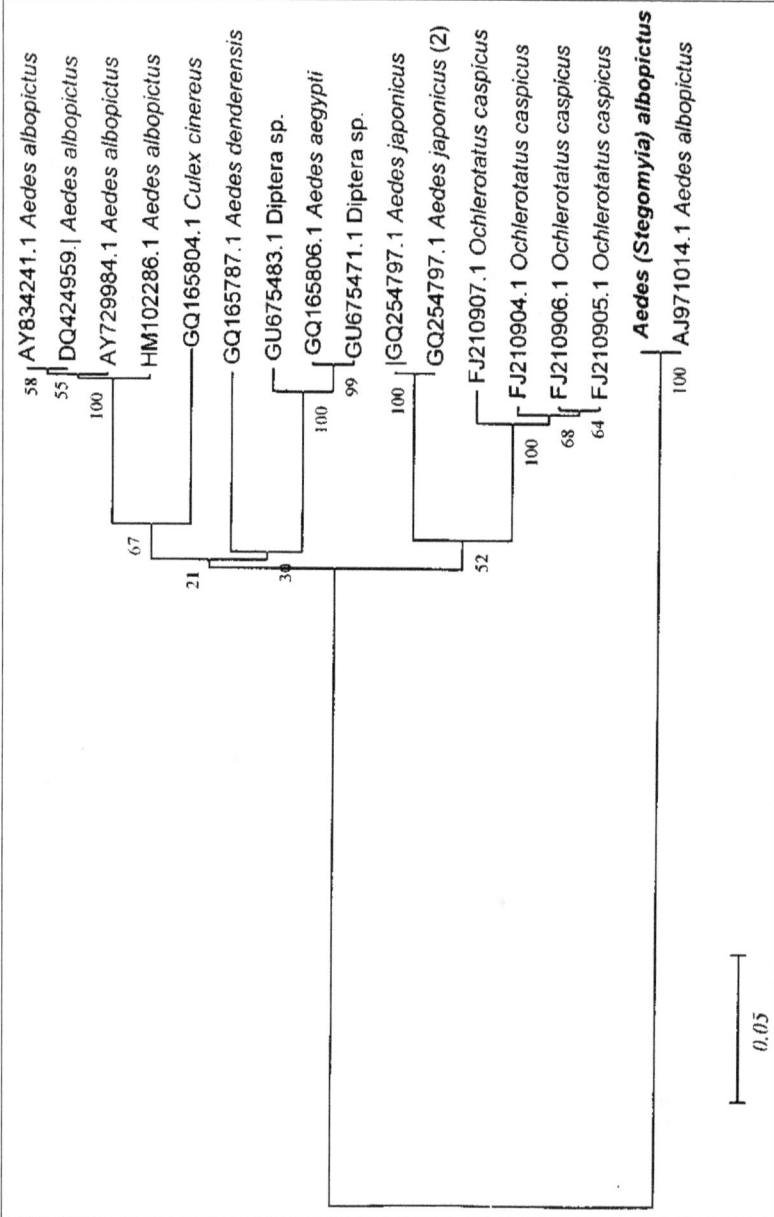

Figure 46: Evolutionary relationships of 17 taxa constructed phyllogenetic tree by using neighbour joining method and MEGA 4 software.

The optimal tree with the sum of branch length = 6.40278786 is shown. There were a total of 404 positions in the final dataset.

6. *Aedes (Finalaya) rajashri* sp. Nov.

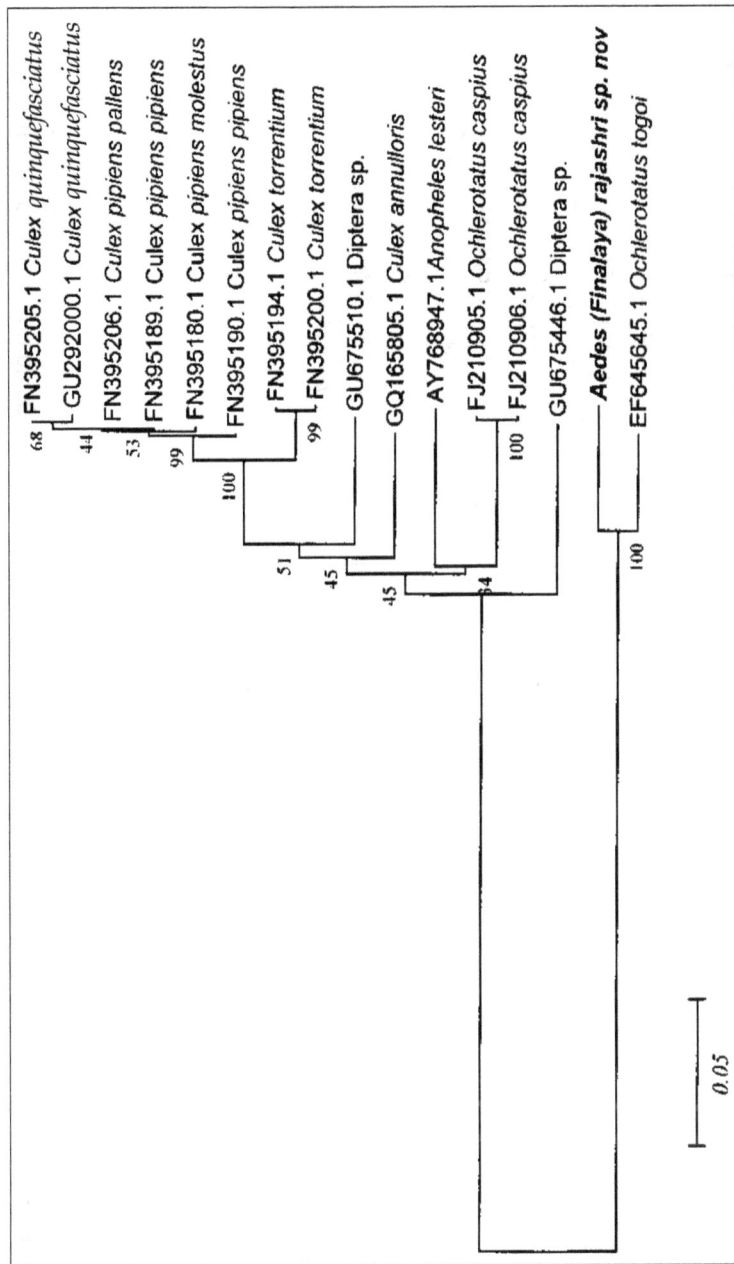

Figure 47: Evolutionary relationships of 16 taxa constructed phylogenetic tree by using neighbour joining method and MEGA 4 software.

The optimal tree with the sum of branch length = 7.19385402 was noticed. There were a total of 443 positions in the final dataset.

7. *Culex* (Culex) *quinquefasciatus* Say, 1823.

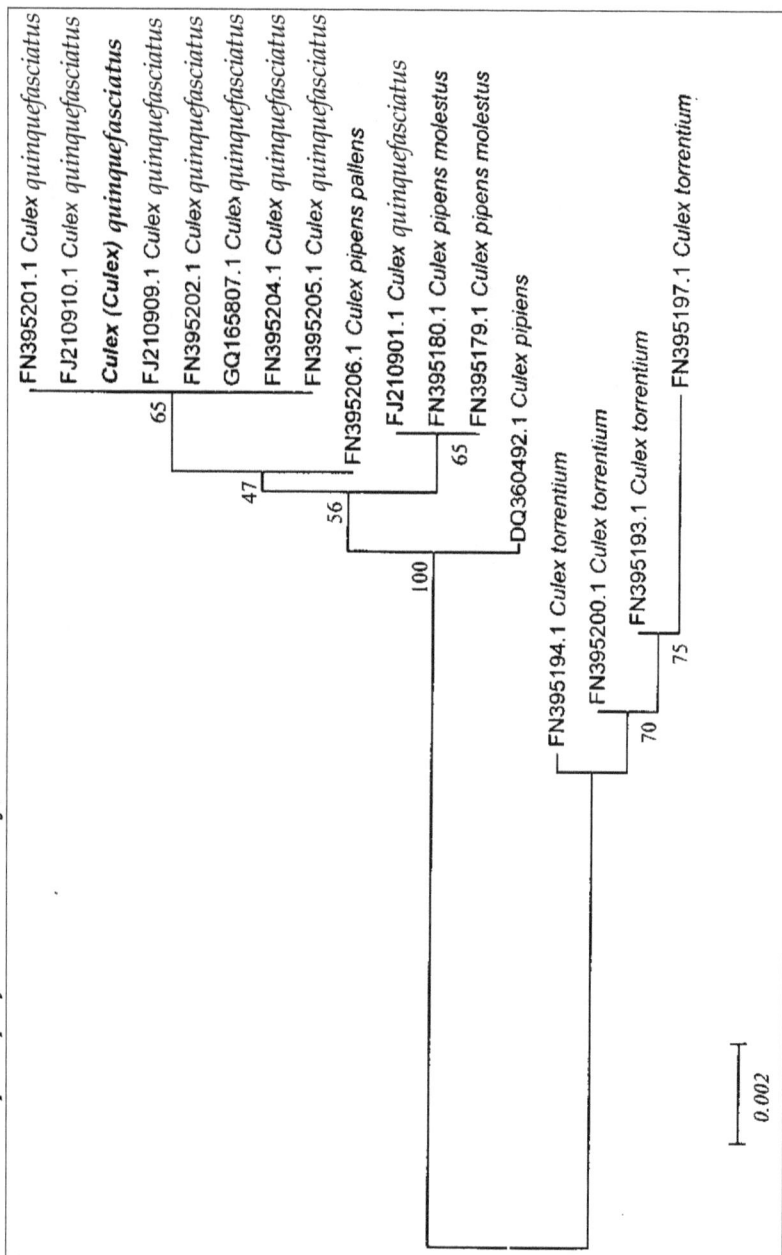

Figure 48: Evolutionary relationships of 17 taxa constructed phyllogenetic tree by using neighbour joining method and MEGA 4 software.

The optimal tree with the sum of branch length = 0.01732766 is shown. There were a total of 635 positions in the final dataset.

8. *Culex (Culex) malakapuri sp. nov.*

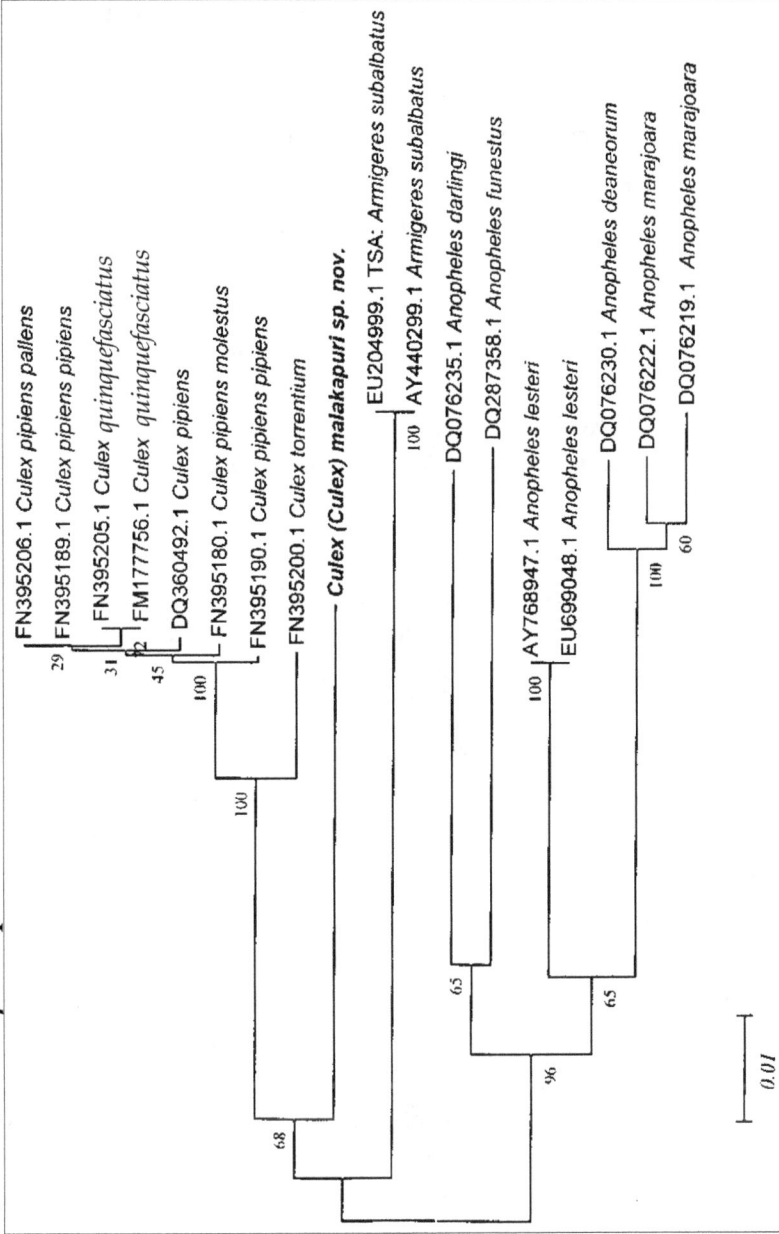

Figure 49: Evolutionary relationships of 18 taxa constructed phyllogenetic tree by using neighbour joining method and MEGA 4 software.

The optimal tree with the sum of branch length = 0.58109042 was noticed. There were a total of 603 positions in the final dataset.

DISCUSSON

The ability of DNA barcodes to identify species reliably, quickly and cost-effectively has particular importance in medical entomology, where molecular approaches to species diagnoses are often of great benefit in the identification of all life stages, from eggs to adults. As Besansky *et al.* (2003) stated "Nowhere is the gap in taxonomic knowledge more urgent than for medically important pathogens and their invertebrate vectors".

PCR amplification of total mosquito DNA with universal 16S rRNA primers yielded a fragment of 500 bp. This product was directly used for sequencing after removal of unincorporated primers and dNTPs. In order to fill up the internal gap, a new set of primers was designed from the conserved regions after alignment of sequences. The primers were LCO (5'-GGTCAACAAATCATAAAGATATTGG-3') and HCO (5' TAAACTTCAGGGTGACCAAAAAATCA-3'). These primers could be used successfully for sequencing 16S rRNA gene fragments of all the 8 mosquito species studied.

The nucleotide base composition of the sequenced rRNA segments showed that – like most other insect mitochondrial rRNA genes (Xiang and Kochar 1991) the mosquito genome too has a high A + T content. It was between 75 per cent and 78 per cent for the species studied by Kambhampati (1995). Studies on nuclear DNA content and chromosome size have shown that *Anopheles* differs significantly from other mosquito species (Rao and Rai 1990). Species of *Anopheles* show a greater divergence among themselves than species of *Culex* or *Aedes*. Classification of insect species is critical for both basic and applied research. The classification based on morphological features poses problems in the case of many groups because of their small size, morphological attributes that change as function of environment and prevalence of biotypes and species that cannot be easily differentiated by morphological criteria. There have been many attempts to use molecular taxonomy techniques to insects and these have yielded valuable results (Xiang and Kochar 1991; Fang 1993; Kambhampati 1995; Tang *et al.*, 1996). Tang *et al.* (1996) analysed a hypervariable region of representative species of three important genera, *Anopheles*, *Aedes* and *Culex*. The gene fragment can be amplified from adults and thus the technique can be used in field studies. Analysis of this and other hypervariable regions may permit an investigation of genetic relatedness of mosquito populations at the subspecific levels. The analysis will be valuable in studies involving molecular taxonomy, particularly for those species that are difficult to identify using morphologic characteristics.In epidemiological research, sequence comparisons of different geographic populations will give estimates of their genetic relatedness and provide information about vector movement.

Gene Sequences, Nucleotide and Amino Acid Diversity and Genetic Distances

The DNA sequences amplified using the primers C1J-1718 and C1N-2191, "500 bp, whereas the sequence with the primer MTFN and MTRN designed was "700 bp. The latter sequence included the former sequences and hence 500 bp was taken for the analysis. This region corresponded to the 5# region of CO-I gene. The sequences were AT rich for the mitochondrial genome, the G and C content being 0.316. The mean genetic distance (K2P) computed for the different species of Culicidae belonging to 15 genera studied was found to be 0.1469. The NJ tree showed 62 species clusters clearly among the sequences studied, thus identifying 62 species among the specimens analyzed. The maximum intra-specific K2P values recorded among the species a cluster was for *Anopheles pallidus* Theobald (0.0184). The inter specific K2P values ranged from 0.0587 (*Anopheles fluviatilis* James s.l. and *Anopheles minimus* Theobald) to 0.2565 [*Verrallina indica* (Theobald) and *Anopheles jamesi* Theobald]. Hence, this study denoted that the K2P genetic distances were 00.02 between different species studied for Culicidae, as cited elsewhere for other group animals (Hebert *et al.*, 2003a,b). However, the K2P value between two very closely related species, *Ochlerotatus swardi* (Reinert) and *Ochlerotatus portonovoensis* (Tewari and Hiriyan), was only 0.0043; thus, they were not identified as separate species. CO-I gene was found to be very conserved as described previously (Knowlton and Weigt 1998), the deduced amino acid variability was only 0.0329. This value indicated the utility of this gene for taxonomy of Culicidae, because it provides ample nucleotide variability toward species identification and its conserved nature for higher orders of taxonomic strata (Hebert *et al.*, 2003a). The nucleotide diversity between the specimen identified as species A and species B was 11.3 per cent, indicating them to be very distinct from each other. The DNA barcode approach could distinguish members of sibling species complexes in insects as reported elsewhere (Hebert *et al.*, 2004a). Unfortunately, the issue of sibling species has given little importance by workers not only to. *An. subpictus* complex but also for sibling species complexes such as *An. culicifacies, An. fluviatilis,* and *An. minimus.* We propose to deal with this aspect in the future. The study of Hebert *et al.* (2004a) indicates the utility of using single- gene sequences (5# region of mitochondrial cytochrome oxidase subunit one gene), toward identification of the mosquito species. The NJ tree computed was in general agreement with the taxonomy based on morphology as reported previously (Hebert *et al.*, 2003a,b. 2004a; Hajibabaei *et al.*, 2006).

Shouche and Patole (2000) studied the sequence analysis of seven mosquito species from India namely *Culex tritaeniorhynchus, Culex bitaeniorhynchus, Anopheles stephensi, Anopheles quadrimaculatus, Aedes aegypti, Aedes w-albus, Aedes albopictus, Culex quinquefasciatus*.

Pradeepkumar *et al.* (2007) separated 111 species belonging to 15 genera of Indian mosquitoes by DNA barcodes. DNA barcode approach based on DNA sequences of mitochondrial cytochrome oxidase gene identified 62 species. The mean genetic distance (K2P) computed for the different species of Culicidae belonging to 15 genera studied was found to be 0.1469. The NJ tree showed 62 species clusters clearly among the sequences. (Pradeepkumar *et al.*, 2007).

Singh *et al.* (2004 a and b) studied the differentiation of members of *Anopheles culicifacies* complex and *An. fluviatilis* complex by an allele specific polymerase chain reaction assay. *Anopheles culicifacies* and *An. fluviatilis* are the a principal malaria vectors in India and *Anopheles culicifacies* is a complex of five cryptic species which are morphological indistinguishable and *An. fluviatilis* is complex of 3 cryptic species. An allele specific polymerase chain reaction (ASPCR) assay targeted to the D3 domain of 285 ribosomal DNA.

The molecular phylogeny of *Anopheles barbirostris* sub group was studied by Claudia *et al.* (2008). The barbirostris subgroup of genus *Anopheles* includes six mosquito species that are almost identical in adult morphology, but differ in their roles in the transmission of malaria and filariasis within Southeast Asia. A 754 bp CO-I mitochondrial gene fragment was sequenced from 136 specimens and the rDNA ITS2 region from 51 specimens. Cywinska *et al.* (2006) identified 37 Canadian mosquitoes by using a short fragment of mtDNA from the CO-I barcodes.

All members of the *Anopheles culicifacies* complex have been identified by Geeta Goswami *et al.* (2000) using allele specific polymerase chain reaction. Polytene chromosome examination has been the only method available for differentiating four species namely A, B, C, D. According to Geeta Goswami *et al.* (2000) the application of explicit methods of phylogenetic analysis is revealing weaknesses in the traditional classification of mosquitoes, but there is strength in intuitive interpretation because the explicit methodology often confirms the monophyly of mosquito taxonomic groups that are diagnosed by unique combinations of characters. The principal problem, then, is not in recognizing monophyletic groups, but in deciding which taxonomic ranks should be assigned to such taxa once their phylogenetic relationships have been established.

However, it is expected that an effective DNA-based identification system should satisfy three conditions: (a) it must be possible to recover the target DNA from all species; (b) the sequence information must be easily analysed, and (c) the information content of the target sequence must be sufficient to enable species-level identification. Amplifing CO-I from total genomic DNA, detected only a single nuclear pseudogene (Gettagoswami *et al.*, 2000).

According to Hebert *et al.* (2004) DNA-based species identification systems depend on the ability to distinguish intraspecific from interspecific variation. CO-I sequence differences among congeneric species were, on average, almost 20 times higher than the average differences within species. The average conspecific K2P divergence for mosquito species (0.55 per cent) was slightly higher than those earlier reported for North American birds (0.27 per cent) and moths (0.25 per cent) (Hebert *et al.*, 2003a). However, this difference reflects detection of deep intraspecific divergence in two taxa (*Ae. fitchii, Ae. abserratus*), instances that may indicate overlooked sibling species. Interestingly, two morphologically distinctive subspecies, *Ae. vexans vexans* and *Ae. vexans nipponi* (dissimilar coloration of scales on the abdominal sternites) show barcode congruence. The latter subspecies was brought to the U.S.A. in 1999, probably from Korea. Hebert *et al.* (2004) detected one case of low interspecific sequence divergence, involving the *Ae. fitchii/Ae. grossbecki* complex. *Ae. grossbecki* is rare in Ontario, although common in nearby northwestern Ohio (Venard and Mead, 1953). Adults of this species were collected from the Windsor London area, its first recorded presence was in Canada (Helson *et al.*, 1978), and specimens of *Ae. fitchii* were collected further to the northeast. The latter individuals showed morphological evidence of hybridization in that two types of scales, were present on the wings of single individuals: large triangular scales, typical of *Ae. grossbecki*, were observed on the anterior half of their wings and elongate scales, typical of *Ae. fitchii*, occurred posteriorly. In cases such as this, indicating possible hybridization, as well as in those cases characterized by incomplete sorting of mitochondrial lineages.

The examination of faster evolving mitochondrial genes, such as the control region or ND4, as well as analysis of nuclear regions, such as internal transcribed spacers (ITS), may aid in establishing species boundaries in at least some of the cases that cannot be resolved though CO1. The effective application of DNA sequence data to molecular diagnostics depends on patterns of nucleotide substitution and the rate of variation among sites (Blouin *et al.*, 1998). The CO-I region in mosquitoes is characterized by a high rate of transitional saturation along the sequence divergence axis, particularly at silent sites. Its saturation begins to level off at around 7.5 per cent sequence divergence, suggesting caution in the interpretation of pairwise comparisons at the congeneric and intergeneric levels, unless silent sites are excluded from analysis. The mosquito fauna of eastern Canada provide an early indication of the patterns of CO-I sequence divergence within and among species, GeneBank sequences for *Anopheles earlei* Vargas from South Africa and *Anopheles quadrimaculatus* Say from the U.S.A. grouped closely with other individuals of their species from Ontario. Moreover, other GeneBank sequences from 'exotic' species, in the genera *Culex*, *Culiseta* and *Anopheles*, grouped with allied taxa in our NJ analysis, but formed distinct, tight sequence clusters. Congeneric species will regularly show sequence divergences in the CO-I region averaging ~ 10 per cent and that divergence values for conspecific individuals will usually fall below 0.5 per cent. In brief the first CO-I barcodes established their effectiveness in discriminating species of mosquitoes recognized through prior taxonomic work in Canada. Specimens of single species formed barcode clusters with tight cohesion that were usually clearly distinct from those of allied species. Sequence divergences were, on average, nearly 20 times higher for congeneric species than for members of a species. Finally it is concluded that an efforts has should be launched to gather DNA barcodes for all known mosquito species, for full evaluation of the effectiveness of DNA barcoding for members of the family Culicidae.

Cywinska *et al.* (2006) identified 37 species of Canadian mosquitoes with the help of DNA barcodes. A short fragment of mt DNA from the Cytochrome C oxidase I (CO-I) region was used to provide the first CO-I barcodes for 37 species of Canadian mosquitoes (Diptera : Cylicidae) from the provinces Ontario and New Bruuswick. Sequence variation was analysed in a 617-bp fragment from the 5' end of the CO-I region. Sequences of each mosquito species formed barcode clusters clearly distinct from those of allied species. CO-I sequence divergences were, on average, nearly 20 times higher for congeneric species than for members of a species. Divergences between congeneric species averaged 10.4 per cent (range 0.2 - 17.2 per cent), whereas those for conspecific individuals averaged 0.5 per cent (range 0.0 - 3.9 per cent). Traditional morphology - based taxonomic procedures are time consuming and not always sufficient for identification to the species level. Therefore, a multidisciplinary approach to taxonomy that includes morphological, molecular and distributional data is essential. Mitchell *et al.* (2002) studied higher-level phylogeny of mosquitoes for mtDNA data support placement of *Toxorhynchites* (Diptera : Culicidae). They assessed the potential of complete coding sequences of the mitochondrial Cytochrome oxidase genes (CO-I and CO-II) and the intervening + RNA - Leucine gene for higher level systematics for phylogenetic affinities of *Toxorhynchites*. On the basis of mitochondrial gene they concluded that the above genus should be placed under the subfamily Culicinae and not under the subfamily Toxorhynchitinae under which only single, above genus is kept. Recent classification of Culicidae recognize three subfamilies *viz.*, Anophelinae with 3 genera, Culicinae with 34 genera and Toxorhynchitinae with a single genus *Toxorhynchites*.

Tamura *et al.* (2004) visualized prospects for inferring very large phylogenies by using the neighbour - joining method. Current efforts to reconstruct the tree of life and histories of multigene families demand the inference of phylogenies consisting of thousands of gene sequences. For such a large data sets even a moderate exploration of the tree space needed to identify the optimal tree is virtually impossible. The neighbour-joining (NJ) method is frequently used because of its demonstrated accuracy for smaller data sets and its computational speed. A likelihood method was developed by these workers for simultaneous estimation of all pairwise distances by using biologically realistic models of nucleotide substitution. This method corrects 60 per cent of NJ tree errors. Hence, it has bright prospects for application in phylogenitic studies and for generating initial evolutionary hypothesis for very large species and multigene family trees. In the present study neighbourhood joining method was used for phylogenetic studies of mosquitoes. The results indicated that there was shifting trend in previously identified species of mosquitoes at molecular level. The present work will be useful for correct identity and availability of mosquitoes in Western Maharashtra, India.

Joseph (1985) studied confidence limits on phylogenies by using the bootstrap method. According to Joseph (1985) recently developed statistical method "bootstrap" can be used to place confidence intervals on phylogenies. It involves resampling points from one's own data, with replacement, to create a series of bootstrap samples of the same size as the original data. Each of these analyzed, and the variation among the resulting estimates taken to indicate the size of the error involved in making estimates from the original data. In the case of phylogenies, it is argued that the proper method of resampling is to keep all of the original species while sampling characters with replacement, under the assumption that the characters have been independently drawn by the systematist and have evolved independently. Majority rule consensus trees can be used to construct a phylogy showing all of the inferred monophyletic groups that occurred in a majority of the bootstrap samples. If a group shows up 95 per cent of the time or more, the evidence for it is taken to be statistically significant. Existing computer programs can be used to analyse different bootstrap samples by using weights on the characters, the weight of a character being how many times it was drawn in bootstrap sampling becomes unnecessary; the bootstrap method would show significant evidence for group if it is defined by three or more characters. Review indicates that very little attension is given on this important method in the world except Cavender (1978, 1981) and Efron (1979). Most methods for inferring phylogenies yield one or few trees and their users rarely go beyond examining the variation among trees that are tried with the best tree under whatever criterion is being emploed. In the present book molecular phylogeny of mosquitoes is made on the basis of barcoding/neighbour-joining method which is helpful for correct identification and rapid cataloguing of species within short time and with less man power. Mosquitoes play the role of vectors of human diseases, hence received special attention of entomologists, health workers and sanitarians of the world.

The tree proposed by Ross (1951) reflected the traditional division of family Culicidae into three sub families. Anophlinae is basal lineage, Toxorhynchitinae intermediate and culicinae is the large derived lineage. The system proposed by Ross (1951) was unchallenged upto 1998. In 1998 Harbach and Kitching examined family culicidae on the cladistic basis. They found Anophelinae as the most basal clade and monophylic. They were also not in opinion for separate sub family for the genus *Toxorhynchites*. Wood and Borkent (1989) and Miller *et al.* (1997) have little doubt for culicidae as monophyletic assemblage. However, Harbach and Kitching (1998) supported the family as monophyletic. Wood and Borkent (1989) listed the premandible of larvae as a synapomorphy of culicidae. Only three of Ross (1951) lineages were supported by Harbach and Kitching (1998) which refers to

1) The monophyly and basal placement of Anophelene.

2) The monophyly of Sabethini and

3) The monophyly of Culicini.

Miller *et al.* (1997) examined the phylogenetic relationships of four mosquito species using 18S and 5.85 rDNA sequences. They placed *Anopheles* in a position of basal to the other species. In a parsimony analysis, *Toxorhynchites* was placed as the sister group of *Aedes* + *Culex* while *Aedes* was the syster group of *Toxorhynchites* + *Culex* in a maximum likelihood tree. The phylogenetic relationship of 13 species of culicidae was tested by using protein-coding white gene by Besansky and Fahey (1997) and accepted traditional 3 subfamilies and nine genera. However, the major disagreement to Aedini + Culicini.

The phylogenetic relationship of 39 species of mosquitoes have been established by Isoe (2000) by vitellogenin gene sequences. Maximum parsimony analyses showed that deeper and terminal relationships were significantly resolved when *Anopheles albimanus* Weid was used to root the trees. VG sequences placed Toxorhynchites in a sister relationship with *Mymomyia*. Hence, *Toxorhynchite* was correlated to place within sub family Culicinae.

Shepared *et al.* (2006) studied molecular phylogenies of mosquitoes and they visualized three traditional subfamilies. They used 18 S rDNA sequences. The *Aedes* (4 spp.) and *Ochlerototus* (15 spp.) formed a distinct clades.

Reinert *et al.* (2006) worked to resolve the relationships and generic placement of species and species groups of uncertain taxonomic position that were previously included in subgenus *Finlaya* by Zavortink (19-12) and placed in subgenera *Protomacleaya* of *Ochleroratus* of genus *Ochlerotatus*. The data were analysed in a total-evidence approach using implied weighting. They obtained the strict consensus of four most parasimonious cladograms. A remarkable congruence in the results have been noticed.

Rey *et al.* (2001) used a 763-bp segment of the mitochondrial CO-I gene to examine the phylogenetic relationships of 14 species of mosquitoes traditionally placed in genus *Aedes* subgenus *Aedes* (1 sp.), *Aedimorphus* (1 sp.), *Ochlerotatus* (11 spp.), and *Stegomyia* (2 spp.). Parsimony and neighbour jointing analyses of the data rooted on *Anopheles claviget* (Meigen) - produced identical trees with all branches supported by bootstrap.

As regards to aedine taxa, Isoe (2000) analysed vitellogenin gene sequences of mosquitoes. He supported the hypothesis of Reinert *et al.* (2004). However, inclusion of *Psorophora* from the *Ochlerotatus* was the disagreement.

Phylogeny of Culicini

The monophyly of culicini was undisputable. This tribe included 789 species of four genera namely *Culex*, *Denocerites*, *Galindomyia* and *Lutzia*.

Culex mosquitoes have been first investigated by Miller *et al.* (1996) on the basis of molecular data. These workers used sequence divergence in the ITS1 and ITS2 regions of rDNA to infer relationships between 14 species representing four sub genera of *Culex*. Neighbour - joining analyses produced a tree that was consonant with Belkin's (1962) notion that Menanoconion be considered as an ancestor of subgenus *Culex*.

Juthayothin (2004) studied molecular phylogeny of *Culex* by using CO-I protein coding gene which showed poor difference between subgenera *Culiciomyia*, *Eumelanomyia* and *Culex*. He also suggested that *Eumelanomyia* is most primitive subgenus.

According to Harbch (2007):

1. Culicidae is monophyletic but deeper relations are unresolved.
2. Anophelinae is monophyletic lineage basal to all other culicidae.
3. Genus *Chagasia* is a monophyletic lineage basal to other Anophelinae
4. Genus *Anopheles* is not monophyletic with respect to be *Bironella*.
5. Subgenera *Kerteszia*, *Nyssorhynches* and *Cellia* are monophyletic and *Kerteszia* and *Myssorhynchus* are sister taxa.
6. Tribes Aedini, culicini sabethini are monophyletic.
7. The genera *Aedes* and *Ochlerotatus* are polyphyletic.
8. With respect to *Denocerites* in genus *Culex* is monophyletic.

The work of Harbch (2007) is very useful for solving problems related to classification and molecular phylogenies. However, much more work is needed on Indian mosquitoes with respect to classification, description, ecology and molecular phylogeny.

Bibliography

Anderson, J. R., Grimstad, P. R., and Severson, D. W. 2001. Chromosomal evolution among six mosquito species (Diptera; Culicidae) based on shared restriction fragment length polymorphism. *Mol. Phylogenet. Evol.* 20:316-321.

Atwal, A.S. 1933. Agricultural pests of India and South east Asia. Kalyani Publ., 1- 289 pp.

Barraud, P.J. 1923a. A revision of the Culicidae mosquitoes of India. Part-V. Further notes on the genera Stegomyia, Theo. and Finlaya Theo. with descriptions of new species. *Indian J. Med. Res.*, 11 : 224-228. 3pls.

Barraud, P.J. 1923b. Two new species of Culex (Diptera, Culicidae) from Assam. *Indian J. Med. Res.* 11 : 507-509.

Barraud, P.J. 1923c. A revision of the culicine mosquitoes of India. *Indian J. Med. Res.* 11: 971-976.

Barraud, P.J. 1924a. A new mosquitoes from Kashmir. *Indian J. Med. Res.* 11 : 967-968.

Barraud, P.J. 1924b. Four new mosquitoes from the Western Himalayas. *Indian J. Med. Res.* 11 : 999-1006. illus.

Barraud, P.J. 1924c. A revision of the culicine mosquitoes of India-XV. The Indian species of the subgenus *Lophoceratomyia* (Theo, Edw.) including one new species. *Indian J. Med. Res.* 12 : 39-45, 2 pls.

Barraud, P.J. 1924d. A revision of the culicine mosquitoes of India. Part XII. Further descriptions of Indian species of *Culex* L. including two new species. *Indian J. Med. Res.* 11 : 1259-1274, 3 pls.

Barraud, P.J. 1924e. A revision of the Culicine mosquitoes of India Part XIII. Further descriptions of Indian species of *Culex* L. including three new species. *Indian J. Med. Res.* 11 : 1275-1282, 2 pls.

Barraud, P.J. 1924f. A revision of the Culicine mosquitoes of India XIV. The Indian species of the subgenus *Culiciomyia* (Theo.) Edw., including one new species. *Indian J. Med. Res.* 12 : 15-22, 1 pls.

Barraud, P.J. 1924g. A revision of the Culicine mosquitoes of India Part XV. The Indian species of the subgenus *Lophoceratomyia* (Theo.) Edw., including two new species. *Indian J. Med. Res.* 12 : 39-46, 2 pls.

Barraud, P.J. 1926. A revision of the Culicine mosquitoes of India. Part XVIII. The Indian species of *Uranotaenia* and *Harpagomyia*, with descriptions of five new species. *Indian J. Med. Res.* 14 : 523-532, 1 pl.

Barraud, P.J. 1927a. A revision of the Culicine mosquitoes of India XIX. The Indian species of *Aedomyia* and *Orthopodomyia* with description two new species. *Ibid.* 14: 523-532.

Barraud, P.J. 1927b. A revision of the Culicine mosquitoes of India Part XX. The Indian species of *Armigers* (including Leicesteria) with descriptions of two new species. *Indian J. Med. Res.* 14 : 533-548, 2 pls.

Barraud, P.J. 1927c. A revision of the culicine mosquitoes of India Part XXI. Description of new species of *Aedimorphus* and *Finlaya* and notes on *Stegomyia albolineata* (Theo.). *Indian J. Med. Res.* 14 : 549-544, illus.

Barraud, P.J. 1927d. A revision of the Culicine mosquitoes of India Part XXII. The Indian species of the genus *Taeniorhynchus* (including *Mansonioides* with a description of one new species. *Indian J. Med. Res.* 14 : 549-554.

Barraud, P.J. 1928. A revision of the Culicine mosquitoes of India. The Indian species of the subgenera *Skusea* and *Aedes*, with description of eight new species and remarks on a new method for identifying the females of the subgenus *Aedes*. *Indian J. Med. Res.* 16 : 357-365, 8 pls.

Baruah, I.; N.G. Das and J. Kalita 2007. Seasonal prevalence of malaria vector in Sonitpur district of Assam. *J. Vect. Borne. Dis.* 44 : 149-153 pp.

Basu, P.C. 1958. A note on malaria and filariasis in Andaman and Nicobar Island. *Bull. Nat. Soc. Ind. Med. Mosq. Dis.*, 6 : 193 – 206.

Beebe N. W., Foley D. H., Cooper R., Bryan J. and Saul A. 1996. DNA probes for the *Anopheles punctulatus* complex. *Am. J. Trp. Med. and Hyg.* 54: 395 - 398.

Belkin J.N. 1962. The Mosquitoes of the South Pacific (Diptera : Culicidae) [sic]. Vols. I and II. University of California Press, Berkeley and Los Angeles.

Besansky N. J. and Fahey, G. T. 1997. Utility of the white gene in estimating phylogenetic relationships among mosquitoes (Diptera : Culicidae). *Mol. Biol. Evol.* 14: 442-454.

Besansky N. J., Krzywinski J., Lehman T., Simrad F., Kern M., Mukabayire O., Fonteinelle D., Toure Y. and Sugnon N. F. 2003. Semipermeable species boundaries between mitochondrial sequence variation. *Proc. Natl. Acad. Sci. USA.* 100 : 10818 – 23.

Besansky, N. J., D. W. Severson, and M. T. Ferdig. 2003. DNA barcoding of parasites and invertebrate disease vectors. *Trends Parasitol.*19: 545-546.

Besansky, N.J., Powell, J.R., Caccone, A., Hamm, D.M., Scott, J.A. and Collins, F.H. 1994. Molecular phylogeny of the *Anopheles gambie* complex suggests genetic introgression between principal malaria vectors. *Proceedings of the National Academy of Science*, USA, 91, 6885–6888.

Bhatia, M.L., B.L. Wattal, M.L. Mannen and N.L. Kerla 1958. Seasonal prevalence of anophelines near Delhi.

Busvine, J.R. 1980. Insect and hygience. The biology and control of insect pests of Medical and domestic importance 521-527 pp.

Chamnarn Apiwathnasorn, 1986. A list of mosquito species in Southeast Asia. Museum and Reference Centre.

Charles R. Vossbrinck, Theodore G. Andreadis, Jiri Vavra and James J. becnel. 2004. Molecular phylogeny and Evolution of mosquito parasitic Microsporidia (Microsporidia : Amblyosporidae). *J. Eukaryot. Microbial,* 51. (1), pp. 88 -95.

Christophers, S.R. 1933. The fauna of British India, including Ceylon and Burma. Taylor and Franas : London 1-360.

Claudia Paredes-Esquivel, Martin J. Donnelly, Ralph E. Harbach, Harold Townson. 2009. Molecular phylogeny of mosquitoes in the Anopheles barbirostris Subgroup reveals cryptic species: Implications for identification of disease vectors. *Molecular phylogenetics and evolution* 50: 141-151.

Crampton J. M. and Hill S. M. 1997. Generation and use of species specific DNA probes for insect vector identification. The molecular Biology of Insect Disease Vectors : a methods manual (ed. By J. M. Crampton, C. B. Beard and C. Louis) pp. 384 – 398. Chapman and Hall, London.

Cywinska, F. F. Hunter and P. D. N. Hebert. 2006. Identifying Canadian mosquito species through DNA barcodes. *Med. and Vet. Ent.* 20, 413-424.

De Meillon, B. 1947. The Anophelini of the Ethiopian geographical region. *Publ. S. Afr. Inst. Med. Res.* 10 : 1-272.

Draft. 2007. National Biodiversity Action Plan, Government of India, Ministry of Environment and forests, pp 1-97.

Edwards, F.W. 1921a. A revision of the mosquitoes of the palaedrctic *Region. Bull. Ent. Res.* 12 : 263-251, illus.

Edwards, F.W. 1921b. A synonymic list of the mosquitoes hitherto recorded from Sweden, with key for determining the genera and species. *Ent. Tidskr.* 42 : 46-52.

Edwards, F.W. 1923. A revision of the Culicine mosquitoes of India Part-III. Notes on certain Indian species of the genus *Finlaya* and descriptions of new species. *Indian J. Med. Res.* 11 : 214-219.

Foley D. H., Beebe N. W., Torres E. and Saul A. 1995. Misidentification of a Philippine malaria vector revealed by allozyme and ribosomal markers. *Am. J. Trp. Med. and Hyg* 54: 46 – 48.

Foley, D.H., Bryan, J.H., Yeates, D. and Saul, A. 1998. Evolution and systematics of Anopheles: insights from a molecular phylogeny of Australasian mosquitoes. *Molecular Phylogenetics and Evolution*, 9, 262–275.

Gillies, M.T. and B. Meillon 1968. The Anophelinae of Africa South of the Sahara. *Publs. S. Afr. Inst. Med. Res.* 54 : 343 pp.

Girhe, B. E. and T. V. Sathe 2001. Incidence of Malaria in Kolhapur district, Maharashtra, *Nat. Sym. Devt. Environ. and Human Cands*, No.3.6, 36 pp.

Girhe, B.E. and T.V. Sathe 2001. On a new species of the genus Aedes *Meign* (Diptera : Culicidae) from India. *J. Adv. Zool.* 22 (1) : 46-47.

Harbach R E. and kitching I. J. 1998. Phylogeny and classification of *Culicidae* (Diptera). *Syst. Entomol.* 23 : 327 – 370.

Hebert, P.D.N., E.H. Penton, J.M. Burns, D.H. Janzen, and W. Hallwachs. 2004a. Ten species in one: DNA barcoding reveals cryptic species in the Neotropical skipper butterfly *Astraptes fulgerator*. *Proc.Natl. Acad. Sci.U.S.A.* 101: 14812-14817.

Hebert, P.D.N., M. Y. Stoeckle, T. S. Zemlak, and C. M. Francis. 2004b. IdentiÞcation of birds through DNA bar codes. *PLoS Biol.* 2: e312.

Hebert, P.D.N., S. Ratnasingham, and J. R. deWaard. 2003b. Barcoding animal life: cytochrome c oxidase subunit 1 divergences among closely related species. *Proc. R. Soc.Lond. B Biol. Sci. (Suppl).* 270: S96-S99.

Juthayothin, T. 2004. Molecular phylogenetic study of culicine mosquitoes using the mitochondrial cytochrome oxidase I gene and the relationships with mosquito-borne flaviviruses. *MSc thesis*, Faculty of Graduate Studies, Mahidol University, Bangkok.

Kanojia, P.C.; P.S. Shetty and G. Geevarghese 2003. A long term study on vector abundance and seasonal prevalence in relation to the occurrence of *Japanese encephalitis* in Gorkhapur district, Uttar Pradesh. *Indian J. Med. Res.* 117 : 104-110.

Korke, V.T. 1932. Observations on Filariasis in some areas in British India, VIII. *Indian Jour. Med. Res.*, 20 : 335-339.

Krzywinski J. and Besansky N.J. 2003. Molecular systematics of Anopheles : from subgenera to subpopulations. *Ann. Rev. Entomol.* 48 : 111 – 139.

Krzywinski J., Wilkerson, R. C., and Besansky N.J. 2001. Evolution of mitochondrial and ribosomal gene sequences in *Anophelinae* (Diptera: Culicidae): implications for phylogeny reconstruction. *Mol. Phylogenet. Evol.* 18: 479-487.

Kumar, A., Black IV, W.C., and Rai, K.S. 1998. An estimate of phylogenetic relationships among culicine mosquitoes using a restriction map of the rDNA cistron. *Insect Molecular Biology*, 7, 367–373.

Lacasse, W.J. and S. Yamaguti, 1950. Mosquito fauna of Japan and Korea, 268 pp. App. I. The female terminalia of the Japanese mosquitoes, 7 pp., App. II. Organization and function of Malaria survey detachments, 213 pp. 34th Edition, illus. off. Surgeon, 8th U.S. Army, Kyoto, Honshu.

Mahendra Jagtap, and T. V. Sathe 2008. Role of Intensified mass surveillance campaign in malaria problematic area of Sangli district. *Perspectives in Animal Ecology and Reproduction* (ISBN10 81-7035-563-X) vol.5, 14-24.

Mattingly, P.F. 1958. The Culicine mosquitoes of the Indomalayan Area. Part-III. Genus Aedes Meigen, subgenera paraedes Edwards, Rhinoskusea Edwards and Cancrdedes Edwards. 61 pp., illus. Brist. Mus. nat. Hist., London.

Mattingly, P.F. 1959. The Culicine mosquitoes of the Indomalayan Area. Part-IV. Genus Aedes Meigen, subgenera skusea Theobald, Diceromyid Theobald, Geoskusea Edwards and Christophersiomyia Barraud 61 pp., illus. *Brist. Mus. nat. His.*, London.

Mattingly, P.F. 1961. The Culicine mosquitoes of the Indomalayan Area. Part-V. Genus Aedes Meigen, subgenera Mucidus Theobald, Ochlerotatus Lynch Arribalzaga and Neomeldniconion Newstead. 62 pp., illus. *Brit. Mus. Nat. Hist.*, London.

Mattingly, P.F. 1965. The Culicinae Mosquitoes of the Indomalayan Area VI. Genus Aedes Meigen Subgenus Stegomyia Theobald (Groups A, B and D).

Maxam A.M. and W. Gilbert 1977. A new method for sequencing DNA. *Proc. Natl. Acad. Sci. USA,* **74**, 560-564.

Miller, B.R., Crabtree, M.B. and Savage, H.M. 1997. Phylogenetic relationships of the Culicomorpha inferred from 18S and 5.8S ribosomal DNA sequences (Diptera: Nematocera). *Insect Molecular Biology,* 6, 105–114.

Moriais I. And severson D. W. 2003. Intraspecific DNA variation in nuclear genes of the mosquito *Aedes aegyptii. Insect Molecular Biology.* 12: 631 – 639.

Murty, S.U.; V.R. Satyakumar; K. Siriram; K. Madhusudhan Rao; T. Gopalsingh; N. Arunachalam and P. Philisamuel 2002b. Seasonal prevalence of *Culex vishnui* subgroup, the major vector of *Japanese encephalitis* in a endemic district of Andhra Pradesh, *India. J. Am. Mosq. Control Assoc.* USA 18(14) : 290-293.

Nagpal, B.N. and V.P. Sharma 1983. Mosquitoes of Andaman Island. *Indian J. Malariol.* 20 : 7-13.

Nora J. Besnsky, David W. Severson and Michael T. Ferdig. 2003. DNA barcoding of parasites and invertebrate disease vectors: what you don't know can hurt you. *Trends in Parasitology,* 19(12):545-546.

Pradeepkumar N, A. R. Rajavel, B. Natrajan and Jambulingam. 2007. DNA barcodes can distinguish species of Indian mosquitoes (Diptera: Culicidae). *J. Med. Ent.* 44(1), 1-7.

Prober, J.M.; G.L. Trainor *et al.,* 1987. A system for rapid DNA sequencing with fluorescent chain - terminating dideoxynucleotides. *Science,* 238, 336-341.

Puri I.M. 1929. A new tree-hole breeding Anopheles from South India. Anopheles sintoni and a revised description of the larva of Anopheles culiciformis Cogil. *Indian J. Med. Res.* 17 : 397, 404.

Rao, P.N. and Rai, K.S. 1990. Genome evolution in the mosquitoes and other closely related members of superfamily Culicoidea. *Hereditas,* 113, 139–144.

Rey, D., Després, L., Schaffner, F. and Meyran, J.-C. 2001. Mapping of resistance to vegetable polyphenols among Aedes taxa (Diptera, Culicidae) on a molecular phylogeny. *Molecular Phylogenetics and Evolution*, 19, 317–325.

Saitou, N., and M. Nei. 1987. The neighbour-joining method: a new method for reconstructing phylogenetic trees. *Mol. Biol. Evol.* 4: 406-425.

Sallum, M.A.M., Schultz, T.R. and Wilkerson, R.C. 2000. Phylogeny of Anophelinae (Diptera Culicidae) based on morphological characters. *Annals of the Entomological Society of America*, 93, 745–775.

Sallum, M.A.M., Schultz, T.R., Foster, P.G., Aronstein, K., Wirtz, R.A. and Wilkerson, R.C. 2002. Phylogeny of Anophelinae (Diptera: Culicidae) based on nuclear ribosomal and mitochondrial DNA sequences. *Systematic Entomology*, 27, 361–382.

Sanger F. 1977. DNA sequencing and chain terminating inhibitors. *Proc. Natl. Acad. Sci. USA*, **74**, 5463-5467.

Santomalazza F., Della Torre A and caccone A. 2004. Short report : a new polymerase chain reaction – restriction fragment length polymorphism method to identify *Anopheles arabiensis* from *An. gambie* and its two molecular forms from degraded DNA templates or museum samples. *Am. J. trop. Med. Hyg.* 70 : 604 – 606.

Sathe, T. V. and Girhe, B. E. 2002. Mosquitoes and Diseases, Daya Publ. House, New Delhi. pp 1-96.

Sathe T. V. 2006. Biological control of mosquitoes. *Proc. Rec. Trends in Malaria Studies*, Pune, pp. 11-17.

Sathe T. V. 2011. Ecology of mosquitoes from Kolhapur district, India. *Int. Nat. J. Pharma and Biosci.*, 2(4): 103-111.

Sathe T. V. and B. E. Girhe 2001. Biodiversity of mosquitoes in Kolhapur district, Maharashtra. *Riv. Di., parassitologia*, 18, LXVII-3, 189-194.

Sathe T. V. and B. E. Girhe 2002. Mosquitoes and Diseases. DPH, New Delhi, 1-122.

Sathe, T. V. and Mahendra Jagtap. 2008. Three decades trend of malaria from Sangli district of Maharashtra, India. *Perspectives in Animal Ecology and Reproduction*, 5, 25-37.

Sathe, T. V. and Mahendra Jagtap. 2009. Tree hole breeding and resting of mosquitoes in Western Ghats, Maharashtra. *J. Exp. Zool., India*, 12(2), 365-367.

Sathe, T. V. and Mahendra Jagtap. 2013. Mosquito Diversity and Control. Daya Pulishing House a Unit of Astral International Pvt Ltd., New Delhi, pp. 1-281.

Sathe, T. V. and Tingare, B. P. 2010. Mosquito Biodiversity, Mang. Publ. Delhi. pp. 1-227.

Sathe, T. V. Asavari Sathe and Mahendra Jagtap. 2010. Mosquito Borne diseases, Mang. Publ. Delhi. pp. 1-342.

Sathe, T.V. and B.P. Tingare 2007. On a new species of the genus Anopheles Meigen (Diptera : Culicidae) from India. *Indian J. Environ. and Ecoplan.*, 14 (1-2) : 61-64.

Sathe, T.V. and Girhe, B.E. 2001. Biodiversity of Mosquitoes (Order : Diptera) in Kolhapur district, Maharashtra. *Riv. Di. Parassitologia*, 18 (LXVII-3), 189-194.

Sen, R.N., R. Rajagopal and Chakraborty 1960. Observations on the seasonal prevalence of adult anophelines near Dhanbad. *Indian J. Malariol.* 14 : 23-54.

Shepard, J.J., Andreadis, T.G. and Vossbrinck, C.R. 2006. Molecular phylogeny and evolutionary relationships among mosquitoes (Diptera: Culicidae) from the northeastern United States based on small subunit ribosomal DNA (18S rDNA) sequences. *Journal of Medical Entomology*, 43, 443–454.

Shouche Yogesh S and Milind S Patole. 2000. Sequence analysis of mitochondrial 16S ribosomal RNA gene fragment from seven mosquito species. *J. Biosci.*, 25. 4:361-366.

Sing,O.P., Chandra, D. Nanda, N., Raghvendra, K., Sunil, S., Sharma, S.K., Dua, V.K. and Subbarao, S.K. 2004. Differentiation of members of the *Anopheles fluviatilis* species complex by an allel-specific polymerase chain reaction based on 28S ribosomal DNA sequences. *Am. J. Trop. Med. Hyg.*, 70 27-32.

Sing,O.P., Goswami, G., Nanda, N., Raghvendra, K.,Chandra, D. and Subbarao, S.K. 2004. An allel-specific polymerase chain reaction assay for the differentiation of members of the *Anopheles culicifacies* complex. *J. Biosci.* 29 275-280.

Stone, A. 1945. A mosquito synonym (Diptera : Culicidae). *Proc. Ent. Soc. Wash.* 47: 38-39.

Swerdlow, H.; S.L. Wu, *et al.*, 1990. Capillary gel electrophoresis for DNA sequencing. Laser induced fluorescence detection with the sheath flow cuvette. *J. Chromatogr.* **516**, 61-67.

Tanaka, K. M. 1979. A revision of the adult and larval mosquitoes of Japan and Korea (Diptera : Culicidae). Contri. *Amer. Entomol. Inst.* 16 : 1-17.

Tewari, S.C., J. Hiriyan and R. Reuben 1987. Survey of the Anopheline fauna of the Western Ghats in Tamil Nadu, India. *Indian J. Malariol.* 24, 21-28.

Theobald, F.V. 1901. A Monograph of the Culicidae or mosquitoes. Vol. 2, viii + 391 pp., illus. London.

Theobald, F.V. 1902. The classification of Anopheles. J. trop. Med. 5 : 181-183, illus.

Theobald, F.V. 1903. A monograph of the Culicidae or mosquitoes. Vol. 3, 359 pp., illus., 17 pls. London.

Tilak, R.K.; K. Duttagutta and A.K. Verma 2006. Vector Databank in the Indian Armey Forces. *MJAFI*, 64(1): 36-39 pp.

Wattal, B.L. and N.L. Kalra 1961. Regionwise pictorial key to the female Indian Anopheles. *Bull. Nat. Sac. Ind. Mal. Mosq. Dis.*, 9 : 85-138.

Weeto M. M., Koekernoer L. L., Karnau L., hunt R. h. and Coetzee M. 2004. Evalution of the species specific PCR assay for the Anopheles funestus group from 11 African countries and Madagascar. Transactions of the Royal Society of Tropical Medicine and Hygiene. 98 : 142 – 147.

Wesson D.M., Porter, C. H., and Collins, F. H. 1992. Sequence and secondary structure comparisons of ITS rDNA in mosquitoes (Diptera: Culicidae). *Mol. Phylogenet. Evol.* 1:253-269.

WHO. 2002. The World Health Report. Reducing risks, Promoting Healthy Life.

Zang J.Z.; Fang, Y. *et al.,* 1995. Use of non-cross linked polyacrylamide for four colour DNA sequencing by capillary electrophoresis separation of fragments upto 640 bases in length in two hours. *Anal. Chem.,* **67,** 4589-4593.

Zavortink, T.J. 1970. The treehole Anopheles of the New world. Contr. Am. ent. Inst. 5(2) : 1-35.

Index

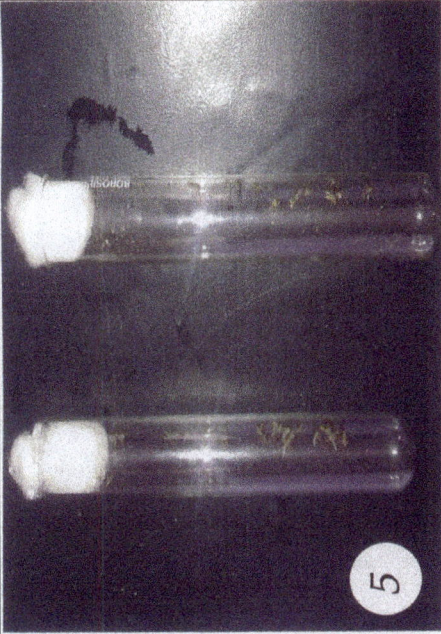

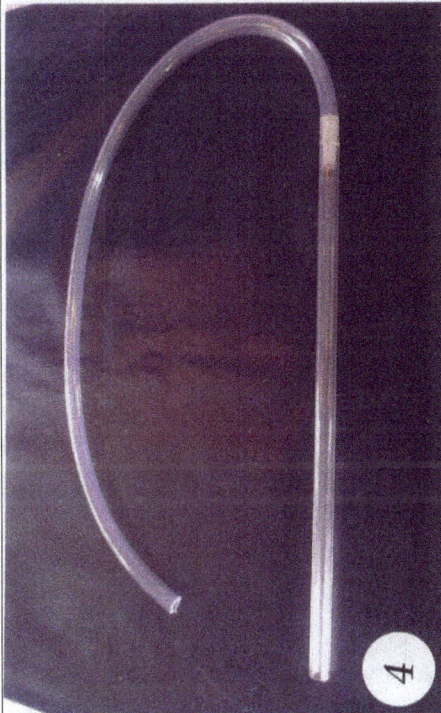

Figure 4: Suction tube; **Figure 5**: Test tubes; **Figure 6**: Torch; **Figure 7**: Mosquito rearing cage.(Page 15)

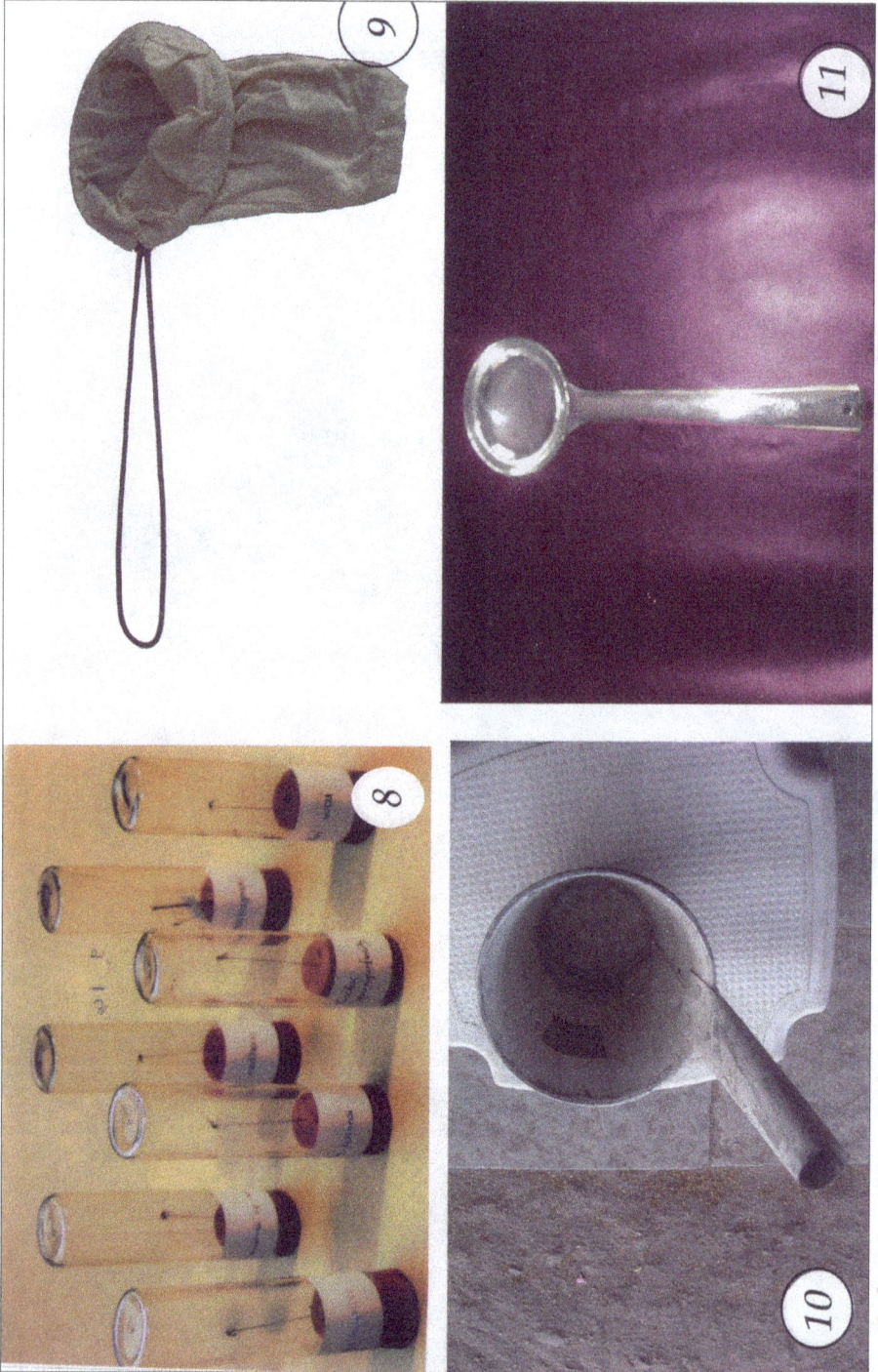

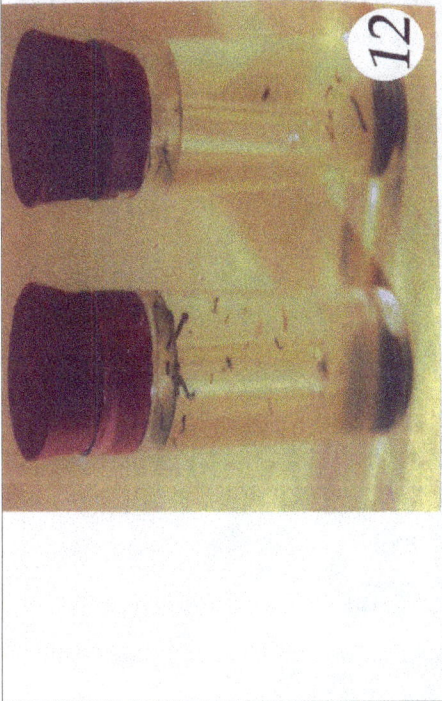

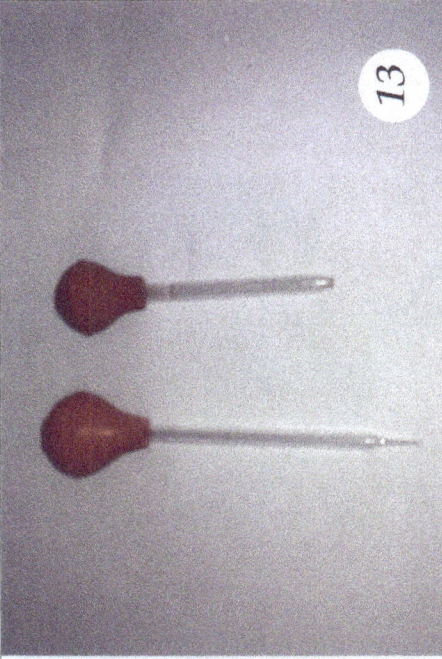

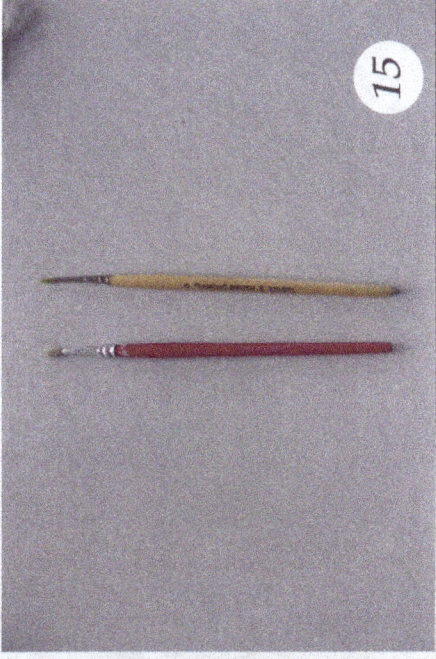

Figure 12: Specimen bottles; **Figure 13**: Dropper; **Figure 14**: Plastic containers; **Figure 15**: Camel brush.(Page 18)

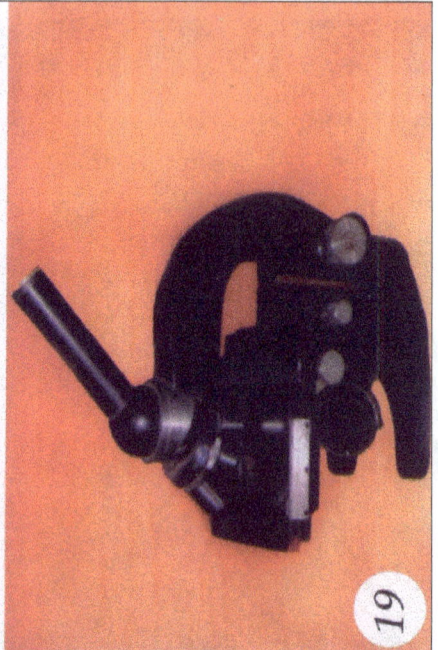

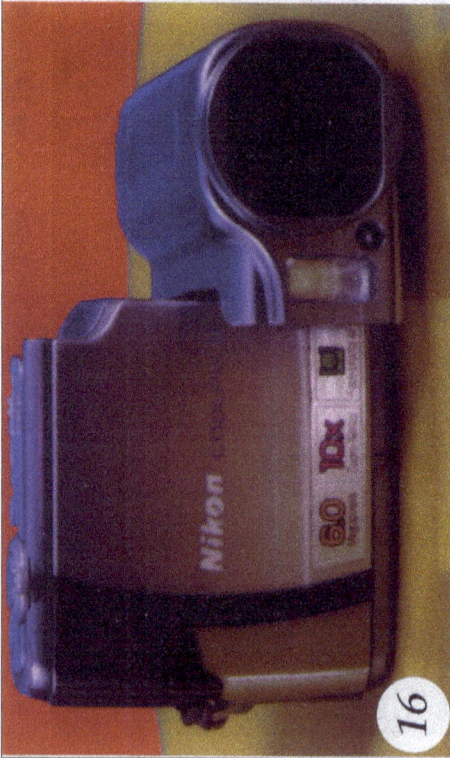

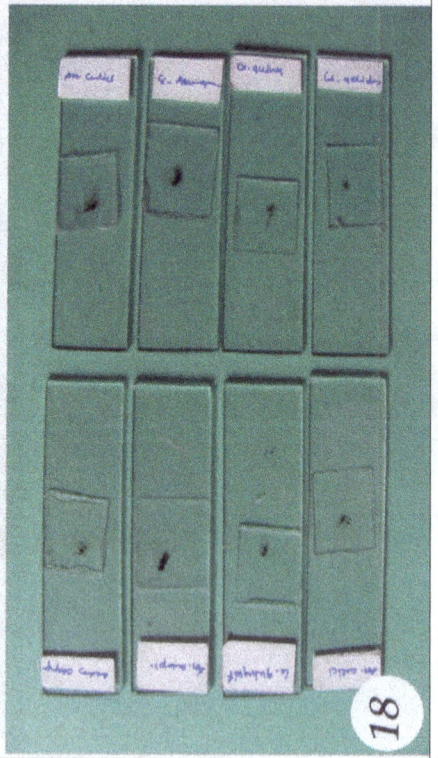

Figure 16: Camera; **Figure 17**: Slide box; **Figure 18**: Slides and coverslips; **Figure 19**: Microscope.(Page 19)

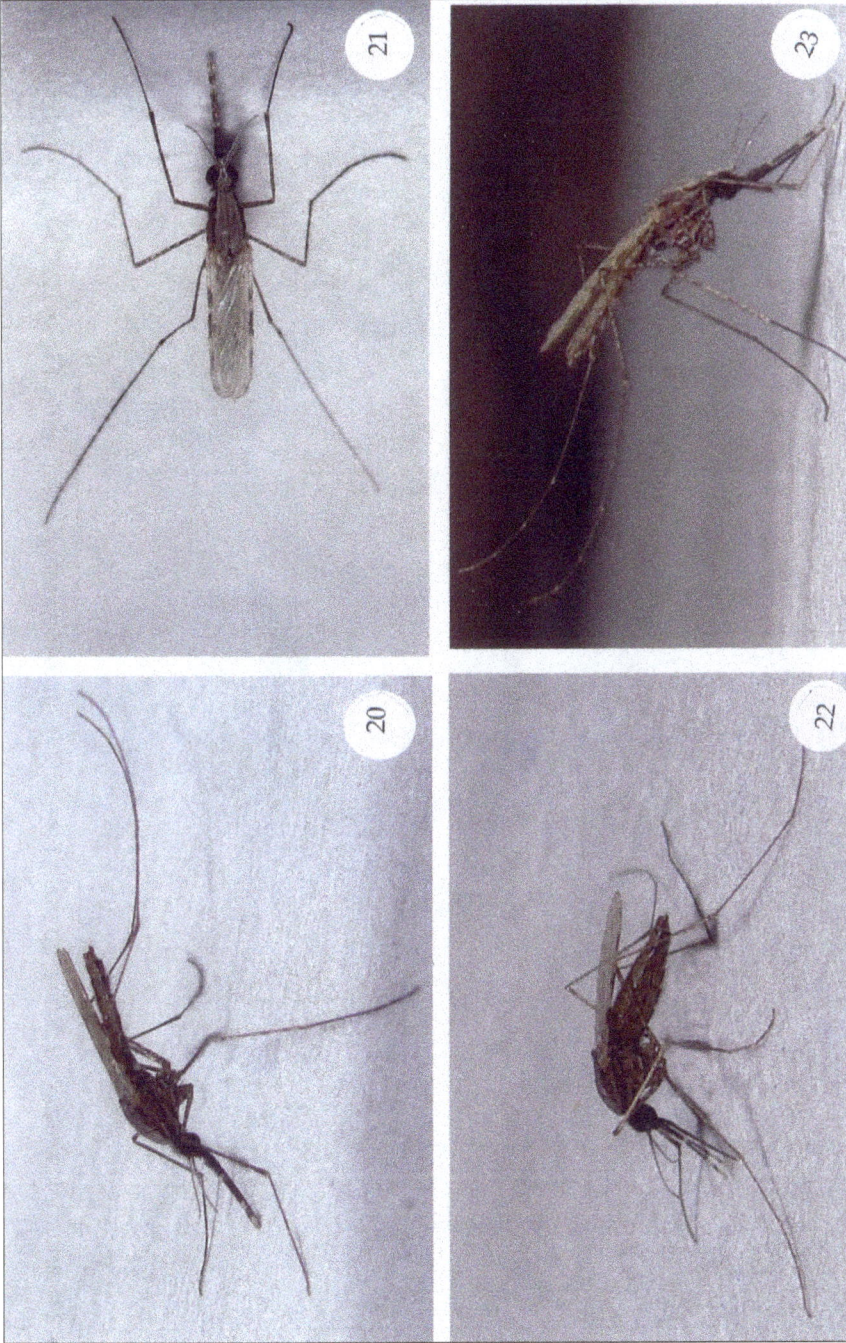

Figure 20: *Anopheles (Cellia) culicifacies;* **Figure 21**: *An. (Cellia) mahabaleshwari sp.nov.;* **Figure 23**: *An. (Cellia) waii sp.nov.; An. (Cellia) karveeri sp.nov. (Page 26)*

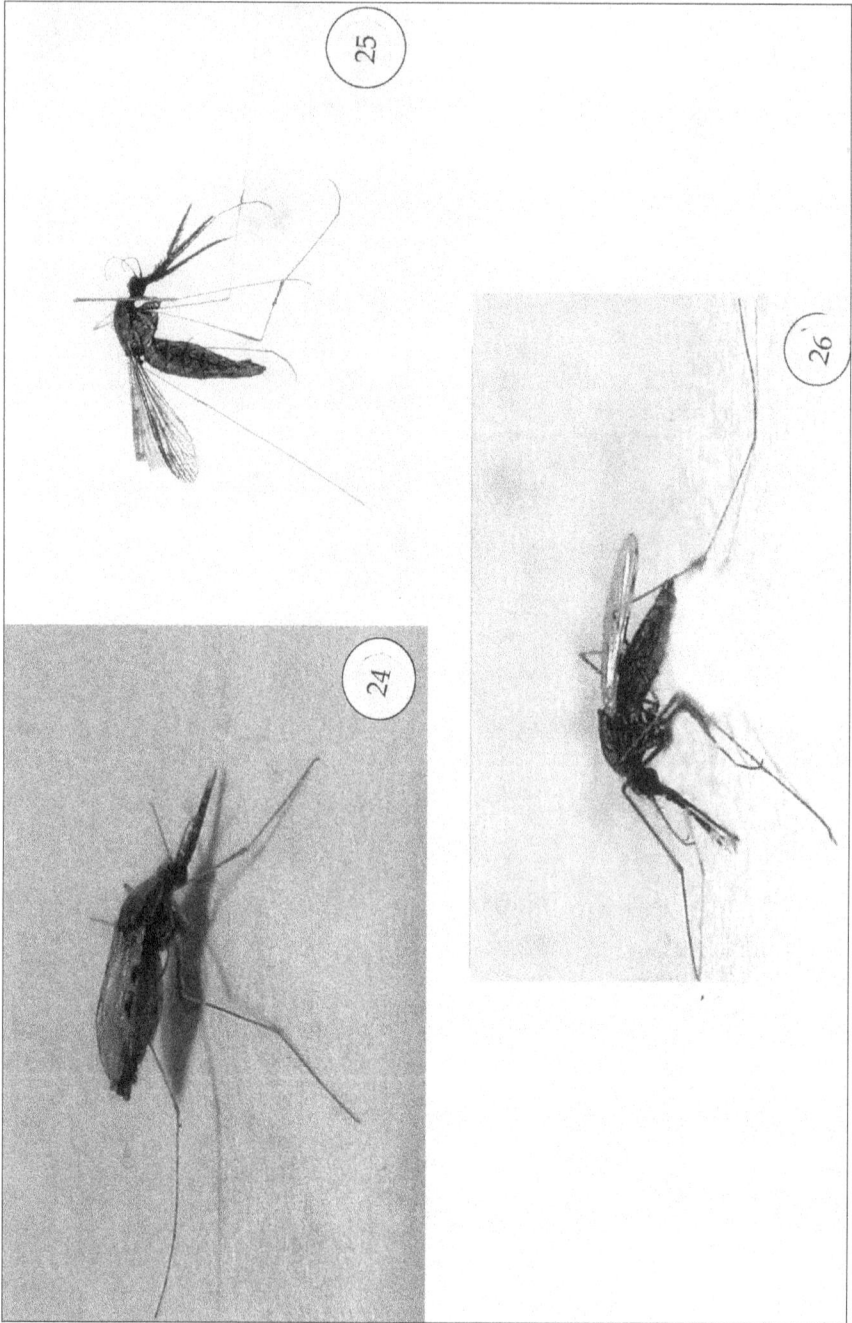

Figure 24: *Anopheles (Cellia) krishnai* sp.nov.; **Figure 25**: *An. (Anopheles) compestris*; **Figure 26**: *An. (Anopheles) kolhapuri* sp.nov.(Page 27)

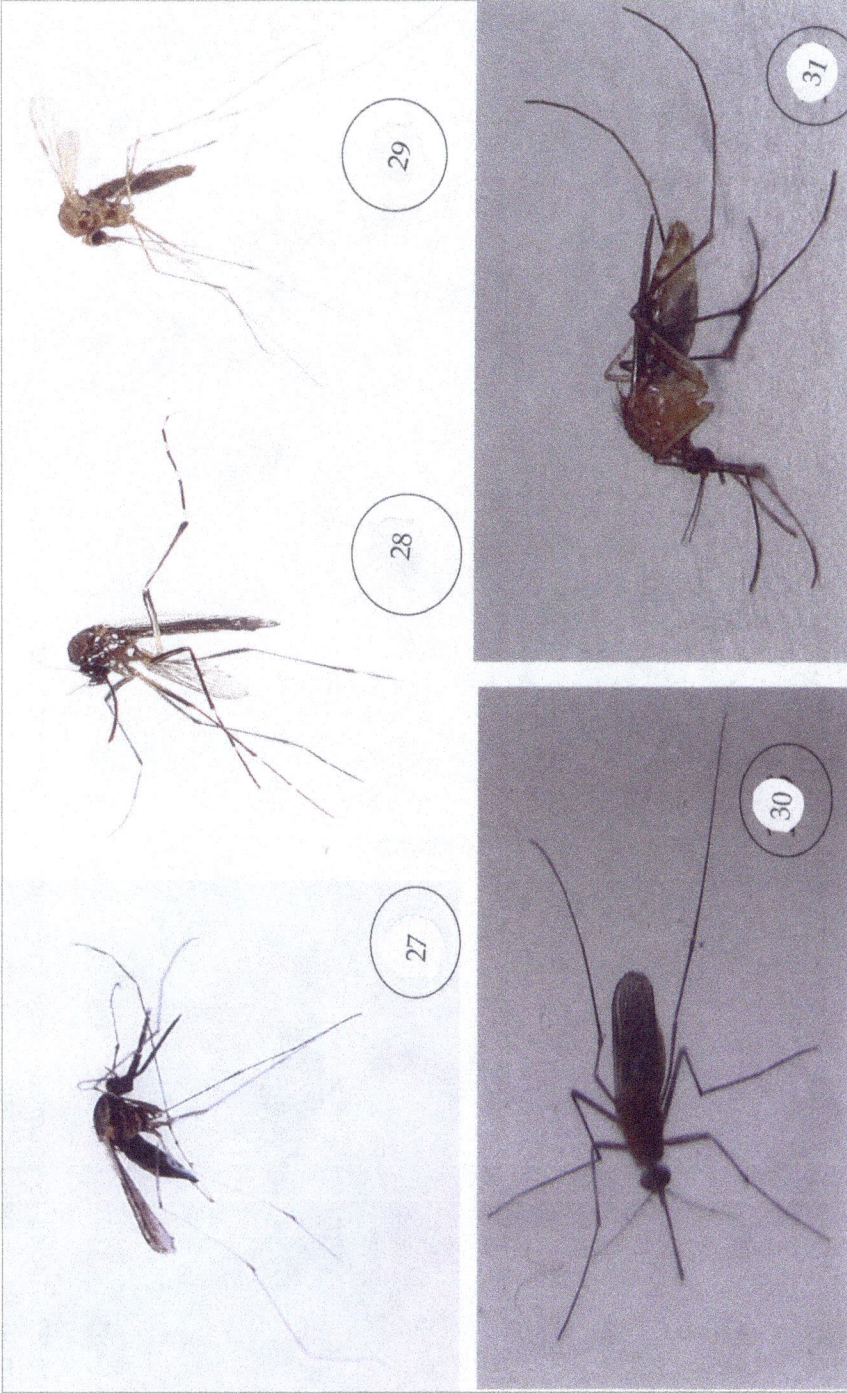

Figure 27: *Culex (Culex) quinquifasciatus*; **Figure 28**: *C. (Culex) malhari* sp.nov.; **Figure 29**: *C. (Culex) malkapuri* sp.nov.; **Figure 30**: *C. (Barraudius) mirajensis* sp.nov.; **Figure 31**: *C. (Barraudius) satarensis* sp.nov.(Page 28)

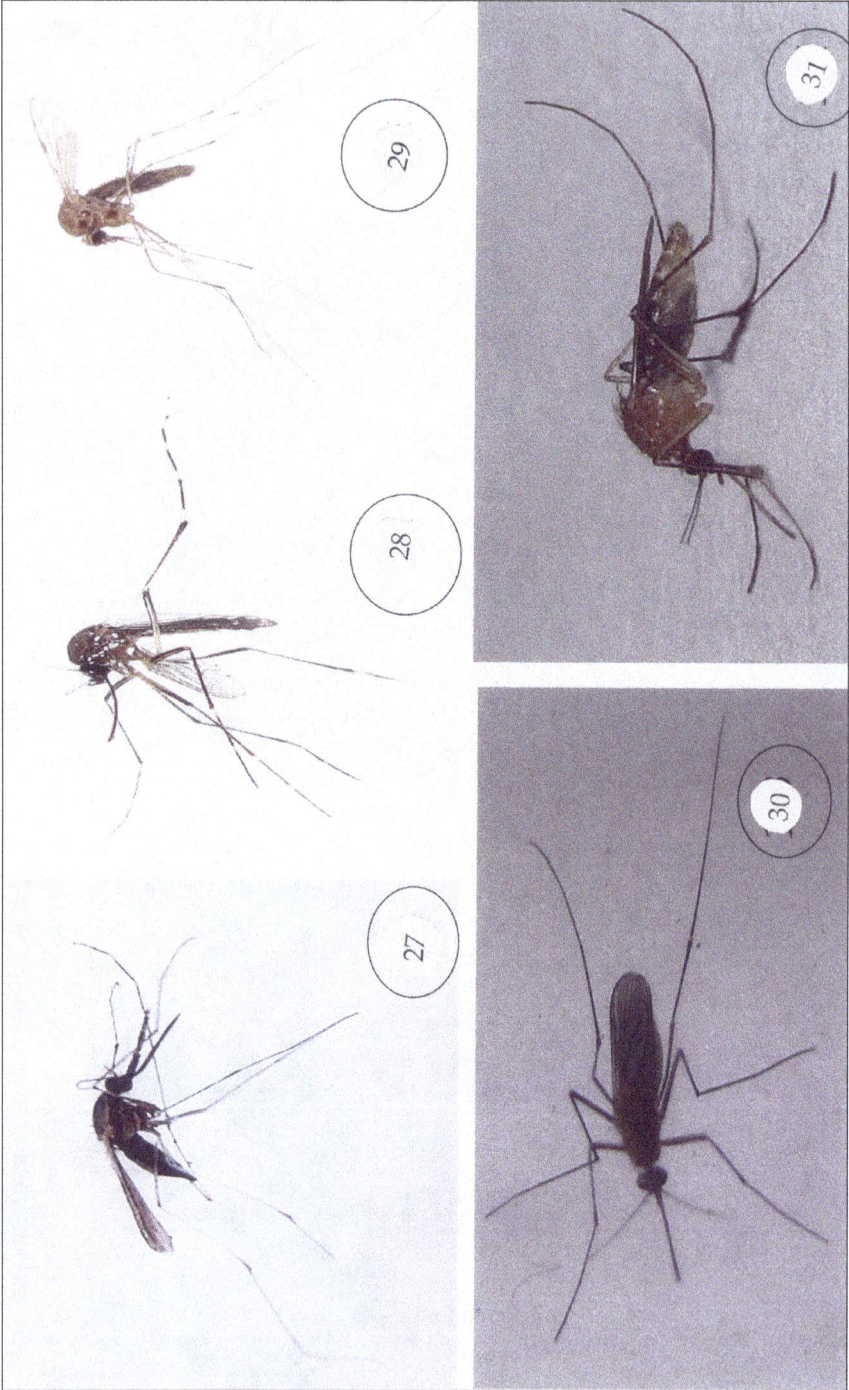

Figure 32: *Aedes (Stegomyia) aegypti*; **Figure 33**: *Ae. (Stegomyia) albopictus*; **Figure 34**: *Ae. (Mucidus) sathei* sp.nov.; **Figure 35**: *Ae. (Finalagya) rajashri* sp.nov. (Page 29)

www.ingramcontent.com/pod-product-compliance
Lightning Source LLC
Chambersburg PA
CBHW050518190326
41458CB00005B/1588